MW00441968

Decision Based Design

Decision Based Design

Vijitashwa Pandey

CRC Press is an imprint of the
Taylor & Francis Group, an **informa** business

CRC Press
Taylor & Francis Group
6000 Broken Sound Parkway NW, Suite 300
Boca Raton, FL 33487-2742

Printed on acid-free paper
Version Date: 20130716

International Standard Book Number-13: 978-1-4398-8232-0 (Hardback)

Library of Congress Cataloging-in-Publication Data

Pandey, Vijitashwa.
Decision based design / Vijitashwa Pandey.
pages cm
Includes bibliographical references and index.
ISBN 978-1-4398-8232-0 (hardcover : alk. paper)
1. Engineering design. 2. Decision making. I. Title.

TA174.P363 2014
620'.0042--dc23 2013025834

Visit the Taylor & Francis Web site at
http://www.taylorandfrancis.com

and the CRC Press Web site at
http://www.crcpress.com

Contents

Preface

This book aims to be the first formal presentation of the central ideas that constitute decision based design (DBD). Engineering education aims to prepare us to solve challenging problems that confront our everyday lives. The connection between curriculum and practice, however, is not always obvious. There are concerted efforts all around the world to bridge this gap and enable the resulting integrated thinking in students, researchers, and practitioners. Through the work of many researchers and educators, DBD has evolved into a discipline in its own right that promises to accomplish this. It views the engineering design process as a set of decisions made by various stakeholders and attempts to model and predict the overall impact of these decisions.

Decision based design in many ways mirrors product design and development. Many fields have developed concurrently to aid product design as well as manufacturing, distribution, and commercialization. Engineering analysis methods have become more and more sophisticated. This sophistication has come at the cost of fuzzy boundaries between different disciplines. A design decision propagates through many levels of engineering and management, and its overall impact is rarely understood well. The field of decision based design aims to train engineers in understanding this interdisciplinary and coupled nature of modern engineering systems.

This book teaches most fundamental concepts encountered in engineering design, such as concept generation, multiattribute decision analysis, reliability engineering, design optimization, simulation, and demand modeling. It can be used in its entirety to teach a course in decision based design, while selected chapters can also be used to cover courses in subdisciplines that make up DBD.

Note to Students

Undergraduate and graduate students can gain immensely from this book. It presents many relevant concepts that you will encounter in the future, whether you choose to stay in academia or move to industry. You will be better equipped to understand an engineering system in its entirety and view yourself as a decision maker. Prerequisites for this book include basic courses in calculus and engineering design. Some knowledge of programming can be helpful in solving exercise problems or finishing projects assigned by the course instructor using this book. Most topics covered in the book are self-sufficient unless you intend to get deeper into a topic, for example, reliability engineering or optimization. At the conclusion of a course utilizing this book, you can be expected to be well versed in core concepts of decision analysis, optimization, product development, simulation, and reliability engineering. If the book is used to teach a course in any of the subdisciplines covered in the book, you are encouraged to browse through the chapters not covered to familiarize yourself with closely related topics.

Note to Instructors

This is a self-sufficient course in decision based design. The course can be supplemented with research papers and relevant books for in-depth analysis of certain areas. The course should include homework that enables students to not only practice concepts, but also think critically about the topic. An involved project can be a very good assistive tool. Students can be asked to apply DBD concepts to the design and development of a product for daily use. In cases where the book is being used for teaching DBD subdisciplines, the following guidelines can be used: Chapters 1, 2, 5, and 6 can be used for courses in optimal design or engineering optimization. Chapters 1, 2, 3, 7, 8, and 9 can be used in a product development course. Chapters 1, 2, 3, 6, and 9 can be used in a stand-alone course in multiattribute decision analysis. Chapters 4 and 6 can supplement a course in reliability engineering.

Any suggestions or comments that may improve the book are always welcome.

Vijitashwa Pandey, PhD

Acknowledgments

This book would not have been possible without the help of many friends and colleagues. I thank Prof. Efstratios Nikolaidis, Prof. Zissimos Mourelatos, and Prof. Deborah Thurston for the direct and indirect guidance provided. I thank friends Cheryl Eash, Akhil Indurti, Ananya Indurti, David Schenk, Danielle Dougherty, Bailey Kretz, Bill Larowe, Monica Majcher, and Dean Mollenkopf, among others, for support and encouragement. The book is also a result of tireless work by Cindy Carelli, Amber Donley, Tara Nieuwesteeg, and Amy Blalock of Taylor & Francis, who were always ready to help when needed. Their patience during the many inordinate delays in manuscript submission is also deeply acknowledged. Finally, I thank my family for their constant support and guidance.

About the Author

Dr. Vijitashwa Pandey received his PhD from the University of Illinois at Urbana–Champaign in 2008. He is an active researcher in mechanical engineering, particularly in the areas of design optimization, decision based design, reliability engineering, and sustainability. His work has appeared in many peer-reviewed journals and conference publications, in addition to a textbook. Dr. Pandey is a strong proponent of interdisciplinary and sustainable efforts in engineering design.

1

Decision Based Design: Introduction

1.1 Is Design a Decision?

Look at the nearest synthetic product around you. Who built it and why? What needs does it satisfy? Why was a particular material chosen? Why is the shape of the product the way it is? What could be done to improve the product? As you ask yourself these questions repeatedly you will realize that at least two (decision-making) entities were involved at some point in time that caused the product to exist: the manufacturer who made the product and the end user who purchased the product (which could be you). The end user had a desire for the functionalities of the product and believed that the resources required to acquire the product were justified. In most cases the resource required is money. The manufacturer fulfilled this need by providing the product at a reasonable price and made some profit in the process.

The interplay between customer needs and their fulfillment by a manufacturer is not always obvious. If we want to design a product, or effect an improvement in one, we would ideally like to be able to answer the question: Which way does the arrow of causality flow in product design—from the customer to the manufacturer (customer demands what the manufacturer makes) or the manufacturer to the customer (manufacturer makes a product and the customer buys it)? A historical perspective on product design is helpful in answering this question. Humanity has been producing tools for millennia. It is, however, only a recent phenomenon that we depend so much on man-made products without necessarily realizing where they come from—most of us were born into a society already dependent on man-made products. Most products that we see are a result of evolution in something simpler. Consider this, nowadays, the need for personal transportation is met by a gasoline-powered automobile. A few centuries ago, this need was met by simpler systems where animals provided the power. Then steam engines and internal combustion engines were built that could increase speed and reliability. Finally, the evolution of the transportation systems culminated into the automobiles we see today. At each step in the evolution of the personal transportation system, either customers demanded something better or the manufacturer provided something extra to stave off competition or

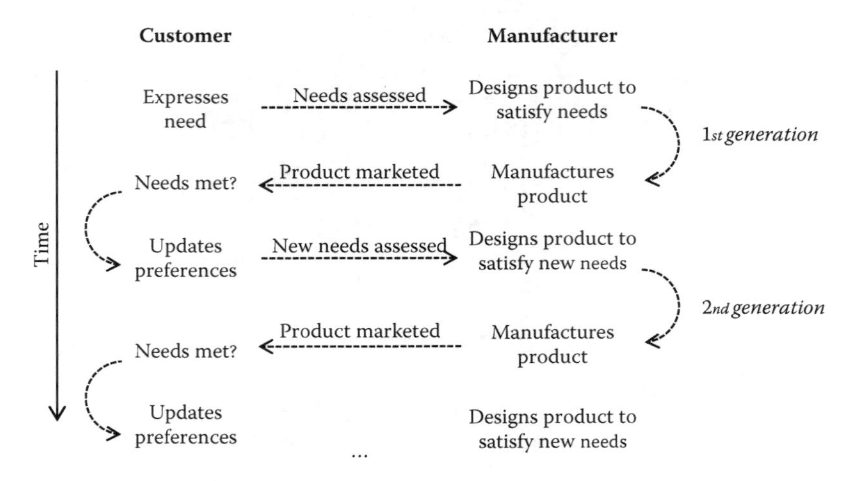

FIGURE 1.1
Evolution of products as a result of evolution of customer preferences.

invite new customers. Something better in personal transportation can mean higher speed, higher acceleration, increased safety, amenities such as car radio, heated seats, and more recently, Bluetooth and navigation systems—or combinations of these. So the causal arrow flowed from the customers to the manufacturer in the sense that they demanded what a car should be like. It also flowed from the manufacturer to the customers in the sense that customers had to choose from whatever the manufacturers offered. However, the causality is not circular; it is separated in time, as Figure 1.1 shows. Customer needs evolve, and so does the manufacturer's design in response to them. Additionally, the manufacturer's design influences customer behavior. We can choose where we want to start and stop in our analysis and, consequently, decide who influences whom.

In this book, the focus is on designing profitable products by making *good engineering decisions*. Consequently, we will be focusing on the manufacturer's perspective, and we will be focusing on products that require engineering. A manufacturer assesses what the customer wants, encodes this information, uses it to design a product, manufactures, and then markets the product, and the cycle repeats unless a new paradigm-changing product (disruptive technology) enters the market or the manufacturer moves on to another product or goes out of business. Since the general methodology within a cycle is the same, we will consider only a single generation in a product evolution cycle, which starts with a customer needs assessment and ends in the product being marketed. This defines the scope of this book. We will be looking at these steps within the framework of decision based design (DBD). We will be breaking up these steps into various technical and nontechnical tasks and identifying what decisions need to be made to accomplish them in the best possible way.

1.1.1 A Manufacturer Makes Decisions

The tasks within a generation of product evolution are very involved. They are planned and performed by individuals. After all, we need to make the decisions of whether to undertake new product development, collect customer preferences information, perform conceptual design, manufacture, and market. Each of these decisions has a huge impact on the success of a product as well as the manufacturer. It is also seen generally that decisions that are made solely on instinct and experience disappoint in practice. Decision based design takes us away from ad hoc decisions and instead gives us a formal method that is, for a given decision situation, repeatable, reproducible, and most importantly, defensible. It gives us a way to think about an involved product development process by breaking it up into small decision-making situations that can then be rigorously analyzed.

Many people follow a very narrow definition of a decision; they simply consider it as the task of selecting between different alternatives. Formal decision analysis and its counterpart in engineering design (DBD) take us a level deeper. In formal decision analysis, a decision is considered an irrevocable allocation of resources. It is considered irrevocable because backtracking to the status quo will cost money or time or both. It is an allocation of resources because they could be used elsewhere. If there were no resources to commit or if we could revoke decisions without losing anything, we haven't really done anything and there is no real need to analyze such a decision. Decision based design considers all the stakeholders as decision makers in their own right. It enables each one of them to understand the task they have been entrusted with and provides a method to make the decision that best serves their needs as well as those of the larger firm. DBD does not question the decision maker's preferences, but only prescribes a way of making a good decision, given these preferences.

Whether it is a new game-changing (disruptive) product or a mature one such as an automobile, the essence of the interplay between customer and manufacturer is always there. While introducing a disruptive product, a manufacturer has to work under tremendous uncertainty. Manufacturers spend significant effort and money trying to determine whether the customers will purchase the product that they are intending to produce. In mature products also, manufacturers need to continuously improve products by adding functionalities so that they stay ahead of the competition. Sometimes, external factors such as legislation effect a change in a product. Manufacturers should also keep an eye out for fundamental advances in technology that can make an established product obsolete. Classic examples of this situation are the electronic calculator replacing the slide rule and digital cameras replacing film cameras.

Once undertaken, a stage comes in the product development process where a design has to be selected and the manufacturing resources committed to it. Clearly, the manufacturer believed that the amount of effort and resources required in manufacturing the product were justified. This justification should be in the form of a clear order from the customer or, in a majority of cases, market studies which have indicated that many customers are willing to purchase the product. How did the manufacturer come to this determination? Did they have perfect information or were they extrapolating based on what they already knew? The manufacturer must have collected information through surveys or some other means. The questions asked in surveys help determine what attributes customers are looking for in a product and at what cost. Market surveys, as most sources of information, involve uncertainty. The manufacturer is always undertaking risk when committing to make a new product. This uncertainty can be formally incorporated if the manufacturer maximizes the expectation of its utility function, as we shall see in Chapter 3. Uncertainties are significant early in the design process, and decisions makers should be cognizant of this fact. Many product development processes are designed to be flexible enough to incorporate new information as uncertainty gets resolved.

Let us now delve deeper into the product itself, which the manufacturer is going to manufacture (the one that maximizes the expectation of the manufacturer's utility). Most products involve many components that need to be designed to work synergistically. Manufacturers require technical competency in making these components (unless it is decided to outsource some of them). The technical competency comes from experience and learning of individuals trained in the relevant subfields of science and engineering. Each of these individuals usually has an assigned task to finish given the targets and constraints. Does an engineer designing a small component also make decisions? Undoubtedly! He or she needs to select the material, from many choices, to make the component. He or she needs to prescribe dimensions, the mechanism of operation, materials, source of power, information flow in cases of electronic components, and so on. Each of these design parameters can be selected in many different ways. The engineer needs to decide which one of these will help the component meet the target requirements in the best possible way, while staying within the design constraints. He or she also needs to understand the effect of uncertainty, which in many cases is irreducible. The job of the engineer does not end at designing the component, though; he or she also needs to inform other members about any new information gained.

There are many system-level problems to be solved as well. Someone needs to coordinate the efforts of various individuals involved in the product design and development process. In most engineering companies a project manager is entrusted with this responsibility. He or she needs to ensure that every individual involved understands his or her tasks, that these tasks start and finish at specified times, and that raw materials and parts needed for manufacture arrive on time. Furthermore, project managers also need to act as liaisons between different individuals and recognize and resolve conflicts as they arise. They also need to have a good sense of where the project stands and if more or less money or resources need to be committed.

A project manager must have the technical background to understand the overall working of the product.

Many times a good design is not good enough. A designer must always look for the *optimal* design. An optimal design is the best design given the constraints. Optimality has a specific connotation in engineering design. It refers to the extremization of a *carefully chosen* function while maintaining its feasibility in terms of cost, safety, and engineering constraints. Obviously there should be agreement on what this function should be and how it is defined. For enterprise-wide problems the most commonly used metric is money, since the manufacturer's main motive is to make money. When solving engineering subproblems, the metrics could be engineering parameters like deflection in a beam, torque produced by a motor, money acts more as a constraint. Sometimes other metrics, such as reliability, commonality index, and design for assembly index, are used as objectives. Doing so, however, is advisable only if they do not conflict with the overarching metric of money or engineering constraints. Furthermore, in each case, a utility function must be assessed over the metric(s) selected.

Every decision maker involved in product design and development works under uncertainty. While uncertainty does not take away the ability to make good decisions, it does affect the ability to predict outcomes with confidence. The sources of uncertainty should be identified and their effects on the design understood. In engineering we refer to uncertainty in the context of a mathematically measurable quantity (random variable) whose realizations can be different each time it is sampled. One would prefer an accurate probability density function for these variables, but many times confidence intervals must suffice. Uncertainty gets into the equation through material properties, manufacturing variability, and uncontrollable factors such as increase in cost of raw materials and market conditions.

1.2 Some Examples of Design Decisions

As we have seen in the previous section, decisions permeate all facets of product design and development. Here we provide some examples of design situations where formal decision-making methodology brings tremendous value:

Enterprise-level decisions:

1. **New vehicle line:** When an automobile company decides to pursue a new line of vehicles, there are numerous variables to consider. What kind of demand exists for a new model? In particular, what are the distinguishing features the new vehicles will have? What level of commonality should exist between the current models and the new vehicles? What technical challenges

should be expected? What legislative requirements (fuel economy, safety) must be incorporated and how?

2. **Nuclear plant siting:** Nuclear plants are considered one of the cleanest sources of energy; however, when they fail, the results are catastrophic. This is why they are usually situated outside high population density areas. However, having them too far from the population can increase the resistive losses in power lines. Also, it is hard to find workers for the power plant in remote areas, particularly if basic amenities such as shopping, hospital, and schools are not nearby.
3. **Offshore oil rig:** Many times significant amounts of oil are trapped under the seabed. Preliminary studies can hint at the presence of oil but cannot guarantee it. Offshore oil rigs are commissioned at locations where tests reveal a high probability of the presence of oil. Even so, a decision must be made, involving vast amounts of money, whether or not to build a rig. The rigs must also be designed considering wind and water surge loads to avoid catastrophic failure.
4. **Offering a new cell phone:** To offer a new cell phone, a manufacturer must understand what functionalities customers are looking for in a cell phone and how much they are willing to pay for it. Manufacturers also want to know what technology to use, e.g., Code Division Multiple Access (CDMA) vs. Global System for Mobile Communications (GSM), and how to compete with competitors.

Engineering decisions:

1. **Material choice:** What material one should use to make a connecting rod for an internal combustion engine is an engineering decision. The material should be rigid, lightweight, and able to maintain its strength even at elevated temperatures inside the engine. Many times special alloys must be developed to increase performance and reliability.
2. **Engine variants:** Market forces sometimes require that a manufacturer make many variants of a product to capture enough demand. For example, a car can be offered in a 2.4 L four-cylinder engine version or a 3.2 L V6 engine version. These differences may require changes in the drivetrain, engine control modules, chassis, and even brakes. The designs should be such that they minimally interfere with the rest of the car, but at the same time provide enough distinction in terms of power to the customer.
3. **Cell phone processor:** Cellular phones today can outperform computers from a decade ago in processing power. The architecture is similar to computers in that there is a central processor

whose speed determines the overall speed and "experience" of the cell phone. However, increased speed in the processor requires compatible memory and video processing units. A faster processor will also drain the battery more rapidly. These factors must be considered when choosing one processor over another.

4. **Conversion of rotary motion to linear:** When converting a circular motion to a linear motion, many different mechanisms, such as the scotch yoke, crank-slider, or rack and pinion, can be used. Each engineering application has different power, speed, and reliability requirements, making one of these implementation options preferable to the others.

1.3 Roadblocks to Engineers Thinking about Product Development

Engineering acts as an interface between science and society. Progressively, it has become highly technical, requiring in-depth understanding of many concepts even while making marginal improvement in products. The result of this is that engineers sometimes tend to focus too much on their engineering subproblems and miss the big picture of profitable product development. While some of the roadblocks to engineers thinking about product development are attributable to engineering becoming highly specialized, others are more systemic and motivational. The engineering curriculum has only recently started acknowledging that in industry, monetary issues often outweigh engineering concerns. A design must be financially feasible (even optimal) to be pursued, regardless of how novel it is. There is also a general motivational aspect attached to engineering. Most engineers enter the field intending to do engineering. They consider financial and managerial aspects of product development to be too boring or nontechnical. This is highly ironical because most successful managers in industry started out as engineers. The good news here is that these roadblocks are being noticed, and most educational institutions offer courses in management and entrepreneurship to engineering students. In Chapter 7, we will revisit these roadblocks to engineers thinking about product development and discuss some additional ones.

1.4 Decision Based Design and Product Development

In this book, we employ the premise that decision based design mirrors product development. We argue that instead of studying the two separately, we should consider an amalgam of the two topics. Product development traditionally is considered to be the sum total of all the activities that together go into making a profitable product. Therefore, the product is the key focal point of the process. Decision based design also gives us a formal way of product realization. However, the contribution of DBD is that it realizes that many decision-making entities are involved that are trying to maximize their own satisfaction, while working toward a common goal of a profitable product. Decision based design therefore puts a human decision maker at the forefront. Not only does this change in perspective better reflect the workings of most companies, but it also provides for a richer understanding of product realization and how it can be accomplished in the best possible way, that is, by making good decisions.

1.5 Why Study Decision Based Design?

As we have discussed in this chapter, products over time have become complex, and their design and manufacture more difficult. More and more features are packed into a product, each of which poses significant technical and nontechnical challenges. Traditionally engineering approached design subproblems separately with very different focuses for each. Many subfields, such as design optimization, reliability engineering, simulation, product development, and decision analysis, were developed to address these needs. Since there is significant overlap between these subfields, a unified framework like decision based design helps by providing context to each design decision. We identified in this chapter, with many examples, that there are decisions being made at each step of product development. As opposed to using intuition, which involves subjectivity (and many concomitant risks), is there a formal methodology for making engineering decisions in the best possible way? Decision based design allows us to do that.

This book stems from a palpable void in the area of engineering design. In the author's time as a graduate student and later as a researcher in the decision based design area, many disparate books needed to be consulted to clarify a simple concept. Books that tended to be somewhat complete in a particular subtopic were so comprehensive that it was hard to find what one was seeking. Discussions with many a new researcher have revealed that they face similar issues; many times they are simply looking for a context in which to put their engineering knowledge. Decision based design provides that context.

There is no shortcut to learning engineering design. Just like the process itself (of engineering design), learning it is also iterative. At every pass one internalizes some concepts, while learning other new ones. This book was intentionally written to be approachable and short enough to be read from cover to cover. Most researchers will find what they are looking for in decision based design and, if not, where to look and, most importantly, what to look for. It will make them comfortable with many important concepts with lucid presentation and solved examples.

1.6 What This Book Does Not Cover

It is important to point out what this book does not cover. This book is not meant to be a compendium of knowledge on engineering design. The book is about what the title suggests, decision based design. It provides a basic framework that helps put accomplishment of engineering tasks in the context of decision making by the stakeholders involved. While enough discussion is provided in the relevant fields of optimization, reliability engineering, simulation, and uncertainty modeling to introduce the reader to these fields, it is not meant to be as exhaustive as dedicated textbooks in these areas would be. The reader hopefully will have developed enough background after reading the book that he or she can decide what aspect of engineering design he or she wants to pursue further. The book is one of the first attempts in presenting these seemingly disparate ideas in an integrated way. The book can be a great starting point in any of these research and practice areas. The interested reader can go about reading dedicated books or publications on the topics covered, if he or she is so interested. This book, however, does provide a comprehensive treatment on engineering design, decision analysis, and product development.

Many basic topics in mathematics and engineering are not covered, and the reader is assumed to be conversant with them. The reader is expected to be comfortable with random variables and their distributions, basic calculus (derivatives and antiderivatives of common functions), probability calculation of compound events, trigonometry, as well as commonly encountered engineering design problems. Where relevant, however, sufficient explanation is provided.

This book also steers clear of a dogmatic approach to decision making in engineering design. We contend that engineering design can never become a pure science. It is repeatable and reproducible for a given set of preferences, in given market conditions, and for a given product type. Even so, there is always room for subjectivity. A rigorous step-by-step approach to decision making in design will assume uniformity in all design problems and the challenges they pose. While seemingly lucrative, it can work flawlessly in one situation and disastrously in others. This is not to say that a proper approach is not beneficial; *it is* for the most part, as the next chapter shows. The rigor and constancy should be more in the way of thinking than in the process itself. The author's work is completed if, after reading the book, the readers feel that every design endeavor is simply the result of a series of informed good decisions made by the stakeholders.

1.7 For Students, Practitioners, and Researchers

The book is primarily a teaching resource, but can be equally valuable for both practitioners and researchers. Practitioners will learn the systematic method to engineering design that has been developed over the years. They will be introduced to the fundamentals of not only the relevant techniques, but also the philosophy of engineering design in the decision-making context. As we have mentioned many times in this chapter, decisions can and must be analyzed formally. A decision made in an ad hoc fashion, more often than not, disappoints in practice. Decision based design helps us think about decisions as allocation of resources, and every decision situation presents an opportunity to minimize the resources consumed and maximize the benefit realized. If practitioners internalize the basic ideas presented here, they will be better prepared to effect positive changes in their organization's best practices.

For seasoned researchers, this book can be a quick resource on engineering design, particularly on the topics with which they are not conversant. The book provides enough background in the topics of engineering design that they can quickly complement their knowledge. A particularly important addition to their repertoire will be decision making under uncertainty, particularly the topics in Chapters 3 and 6. For researchers new to the area of decision making in engineering design, or even just engineering design, this book can be a valuable resource. Unlike a dedicated book on decision analysis, or optimization, this book stays rooted in design. The book provides the most value for students—both graduate and undergraduate. If introduced early to DBD, they will be able to shape their education and learning in a way best suited to their interest, while being fully conversant in formal normative decision making.

Problems and Exercises

1. Comment on the interplay between customers and manufacturers when designing
   a. A new product
   b. An improved already existing product
2. How does uncertainty affect engineering design decision making?
3. Enumerate different decisions a manufacturing firm has to make for successful realization of a product.
4. How are enterprise-level decisions different from technical decisions?

5. Do you agree that profit making is the sole purpose of an engineering organization?
6. What is the role of a project manager?
7. Why do you think engineers can make better managers?
8. Why should one study decision based design?
9. Do you agree that decision based design mirrors product development? Provide explanations.
10. What value does decision based design bring to a fresh engineering student?
11. How will you convince someone to take up decision based design as a topic of research?
12. How will you convince an experienced engineering manager to use decision based design concepts in his or her design?

2

Engineering Design

2.1 Engineering Design toward a Product, System, or Service

Engineering design is pervasive around us. There is hardly anything we see or interact with in real life that is not the result of an engineering endeavor requiring conscious effort. From the design of a simple stapler to the design of a wide-body aircraft, all are results of engineering design. Engineering design has a broader scope than just solving the technical problem of meeting design requirements specified by the customer. It must consider all the extraneous factors: the motivation behind a design, expected deliverables from the design and final attributes of the product, the effect of uncertainty, safety parameters, environmental impacts, market needs, and legislation, among others. The outcome of a successful engineering design process is clear specifications on how to create a product, system, or service that will meet all the deliverables.

Engineering design bridges the gap between science and its successful use in simplifying our lives. Engineering design creates products, systems, and services that employ scientific theories in their functioning. There are, of course, differences in products, systems, and services in the way they are defined. A product is a well-defined physical artifact that does a set of functions usually under end user control. A system is a differentiated collection of engineering artifacts that work together to achieve a wider set of requirements. A service may or may not include a physical artifact but is usually a contract to fulfill engineering duties in exchange for compensation. There are differences in the role of engineers in enabling the three, but we will refer to these as products for the rest of this book.

Central to engineering design are the designers, or human entities who perform the tasks required to turn a design concept into reality. The change in perspective of engineering design being considered more from a designer's point of view is a relatively recent phenomenon. Engineering design up to the Second World War era and even shortly afterwards was still concerned with precisely specifying parameters to manufacture a product. This philosophy changed as engineering systems became more complex and design became truly interdisciplinary. Customer voice became stronger, which led to interaction of groups within organizations that used to be disparate. Product features and long-term reliability had to be accounted for. It also led manufacturers to reconcile the technical with the nontechnical (safety, brand image, sustainability). Commoditization of many technologies asked for efficient manufacturing and marketing methods. There were subject matter experts on a product development team who now had to constantly interact, despite not being able to fully appreciate each other's point of view. The advent of computers and their near ubiquitous use in design, on the one hand, streamlined some aspects of design, but on the other, also brought to the forefront the amorphous nature of a highly complex design process. Clearly a formal methodology of design was needed.

It is important here to make the distinction between design science and design methodology. Design science is a field more concerned with understanding the technical aspects of a design and its effect on the environment (Pahl and Beitz, 1995). Design methodology, building upon design science and being informed by the practice of design in industry, attempts to make a systematic approach to design. It helps designers break down a design problem into constituents and approach them systematically. Making design a systematic method does not constitute making it dogmatic, as Pahl and Beitz point out; it is more a way of giving design a direction. It makes the designer realize that creativity and intuition have to have a direction for successful realization of a product. A systematic approach to engineering design has many tangible and intangible advantages, as we will see throughout this chapter.

As engineering design research started developing a structure, many subdisciplines began to be realized. Coordination of design efforts across various subdisciplines became an issue that needed to be tackled. Design as a field of study in and of itself led to development of many design methodologies, all realizing more or less the central role a human designer plays in providing the motivation for and in coordinating design efforts. Pioneering work was done by Shupe et al. (1988), Mistree et al. (1990), Suh (2001), Thurston (1991), Hazelrigg (1998), and Lewis and Mistree (1997) in the eighties and nineties. Engineering design is now universally considered a sequence of informed decisions made by designers. In this approach, a decision is considered the smallest unit of communication between the designers (Lewis and Mistree). Understandably, the designers are expected to make good decisions. This realization, combined with the fact that there were a lot of concurrent breakthroughs being made in the field of normative decision analysis, gave rise to the field of *decision based design*. Thurston then proposed that given the mathematical model of an engineering system, the only best design is the one that maximizes the designer's expected utility. This idea welcomed normative decision analysis into engineering design.

2.1.1 Decision Analysis and Engineering Design

A decision is defined as an irrevocable allocation of resources. As it turns out, there are many resource allocation decisions in an engineering design process. The study of how to make good defensible decisions is called decision analysis. The term *decision analysis* was coined by Ronald Howard (1966). Prior to that, an axiomatic approach to making decisions was proposed by von Neumann and Morgenstern (1944). Their approach provided for the expected utility criterion to making decisions. The axioms have been presented in slightly different ways by different authors, for example, Savage (1954), Keeney and Raiffa (1994), Nikolaidis, Mourelatos, and Pandey (2011), and Howard (2007). The five rules of actional thought, because of Howard, capture all the necessary elements in making decisions, and we will be using them in this book, particularly in Chapter 3. Decision analysis is touted as a normative field; i.e., decision analysis does not try to address or even model what a decision maker would do, but tells us what a designer should do, given his or her preferences. Aside from decision analysis, heuristics such as fuzzy logic (Zadeh, 1996) and analytical hierarchy programming (Saaty, 1990) were also proposed and used successfully in engineering design. In this book, however, we will focus only on decision analysis because it does not depend on ad hoc performance functions but allows us to perform value of information studies in a mathematically consistent manner.

Use of normative decision analysis is now a well-accepted notion in engineering design, as evidenced by the number of researchers involved and publications, as well as examples of successful practical applications. It provides a systematic way of understanding and encoding decision maker preferences under uncertainty into a utility function, which can rank alternatives in the order of desirability. There are challenges to assessing mathematical forms of utility functions, though. The effort required from the decision maker as well as from the facilitator (the person helping the decision maker) is sometimes substantial. This leads to a lack of enthusiasm from designers in providing sincere answers to lottery questions commonly used to assess utility functions. This has been cited as the reason for lack of universal acceptance. This issue is further exacerbated when multiattribute utility functions are needed as these are much harder to assess than single-attribute utility functions, unless some independence conditions are met. There are also some misconceptions about decision analysis that keep designers from using it in practice. Some of these stem from lack of understanding of the proper methodology, while others are results of biases from which most decision makers suffer. Decision analysis is not a cure-all. A design that maximizes the expected utility may disappoint in practice if the utility functions or design uncertainties are not carefully assessed. Thurston (2001) talks about the real and misconceived limitations of multiattribute decision analysis in design and highlights that the quality of design decisions made using decision analysis is a direct function of the effort put into it.

Engineering design, like most scientific disciplines, has prescriptive and descriptive aspects. Also as with any other scientific discipline, there are positives and negatives to both, and consequently proponents and opponents. A prescriptive approach looks for a definitive approach to design akin to that proposed by Pahl and Beitz (1995), which is how rational designers would make design decisions. A descriptive approach tries to understand decision making by human designers and tries to predict their behavior. A real-life complex design process should have elements of both, because there is always substantial influence from the human entities involved irrespective of the level of automation and computer-based design decisions. Furthermore, since decisions are almost always made with limited information and by individuals with bounded rationality, we might never get to the point where prescriptive design methods are the norm. A good approach therefore looks for the confluence of prescriptive and descriptive methods. Design decisions under the assumption of bounded rationality are an active area of research (Gurnani and Lewis, 2008).

2.1.2 Arrow of Causality: Design in a Company Setting

In Chapter 1, we discussed how within one market cycle, the arrow of causality for design flows from the customer to the manufacturer; i.e., the customer expresses a need, and the manufacturer responds by designing, manufacturing, and marketing a product that satisfies this need. What happens when we zoom in to the design process within an organization? Regardless of the setting, the arrow of causality in design necessarily flows from one decision maker to another. In a for-profit company setting, a business development manager or a product planning team identifies a market opportunity. A market research is initiated to find out the customer needs and potential market for the product. This information is then presented in a standardized needs statement to a team of designers. Based on the needs statement, a preliminary decision is made to explore the design of a product. Each designer brings expertise from his or her discipline; for example, in the design of an electromechanical switch, mechanical, industrial, and electrical engineers might be needed. Other team members include production managers, manufacturing personal, and quality engineers. After concept generation and selection, detailed design follows (unless it is decided to outsource). Therefore, what causes a designer to start working on a design is usually his or her supervisor telling him or her in clear terms what is needed from him or her. Depending on the problem setting, these designers have specific roles that are ideally aligned with their skills. In short, one decision maker (DM) identifies the need to be fulfilled by a product, identifies the best person (a DM in his or her own right) to fulfill it, and entrusts him or her with the job of designing it in exchange for compensation. As a result, fortunately or unfortunately, almost all engineering design is driven by the monetary needs of a company or the individuals involved.

2.2 Design Methodology

A design methodology is a structured approach to realize a design that meets all the required specifications. We present a short description of the engineering perspectives on design here; in Chapter 7, on product development, a more classic (and expanded) approach will be presented. The type of products to be manufactured determines to a large extent the design effort required and the intricacies of the methodology used. There are established products, new mass-produced products, and new one-off products. Established products are the ones that the manufacturer has been making for a long time. The technical challenges have already been tackled, the manufacturing process is streamlined, and the workforce is trained. Improvements typically required for established products are marginal. They are usually related to the addition of new features, cost reduction, and improving reliability. Complete redesign is rarely undertaken. In fact, established products generally prove to be a hindrance when a fundamental change in products is envisioned. Most companies prefer to launch new product lines afresh when demand for an established product starts to wane. Many times protection of market image becomes critical, and companies have been known to start spin-offs to pursue new, potentially risky product ventures.

New products are sometimes a result of an identified market need, while at other times they are a clear order from a customer for a mass-produced or one-off product. An internal motivation for a new product comes when a company identifies that customers need a product that can perform a given function better than the alternatives available in the market. Alternatively, a company may notice that the products it currently manufactures have lost profitability and new variants need to be developed. The company may also want to counter threats from competitors by staying ahead. New products require the most design effort. When they are designed to meet a projected market need, the uncertainty associated with their marketing success is large. Customer requirements need to be carefully assessed, new technical challenges confronted, manufacturing methods developed or modified, and workers trained. Reliability is also an issue for new products due to many inadvertent design and manufacturing mistakes. As such, new products can make or break a company. This is why new product development requires a structured approach and forms the focus of this book.

External motivation comes from a customer who orders a new product. Clear orders from the customers for new mass-produced or one-off products reduce marketing uncertainty. The orders also provide a clear description of what is needed in a product; therefore, the design uncertainty is also somewhat reduced. There are some critical differences between customer orders for new mass-produced products and those for one-off products. One-off product development projects are typically large projects that companies undertake in a project-based environment. They need a substantially new design and even require development of new technologies. Since large profit margins are often available, one-off products can form a platform for creating core competencies in related product types. The experience gained from such projects can be immense for future similar projects. Ideas for mass production of newly developed technologies can also be tested.

In a proper design methodology, while the detailed procedure may differ for different products and the organizational structure of the firm, the following steps are generally undertaken:

1. Understand problem definition
2. Perform basic feasibility studies
3. Decompose design into subproblems
4. Brainstorm and solve subproblems
5. Integrate solutions to subproblems
6. Revisit steps

The design process begins with a careful understanding of the problem at hand by all the decision makers involved. Various decision makers (project managers, design and manufacturing engineers) get together and discuss the scope of design in this step. Pahl and Beitz (1995) call this step clarification of task. Regardless of whether it is a result of internal motivation or a customer order, the requirements from the product should be clear in the minds of all the decision makers involved in the design process. Ideally, before the decision makers start proposing engineering solutions, they should understand what essential functions the product is required to perform, what engineering solutions already exist, what the budget is, what resources will be required, what the product launch date is, and what manufacturing methods will tentatively be used. Clearly, some of these questions, particularly resources required and manufacturing methods used, are difficult to answer in the initial stages; therefore, educated guesses from experienced engineers are very important.

Before a design project is formally undertaken, the company should perform feasibility analyses. Technical feasibility determines whether the product can be produced within the engineering constraints under which the company operates. A financial feasibility study determines whether the product can be *profitably* manufactured and marketed, if it is technically feasible. Technical and financial feasibility go hand in hand and very often conflict. Usually before an order is accepted or a product design process is launched, a meta-decision is made to determine technical and financial feasibilities of "ballpark" designs. Most researchers agree that companies' prime objective is making money; therefore, a money criterion can be used. A risk-neutral decision will involve enumerating different possible designs and their expected costs to manufacture. These designs are, by definition, very abstract at this stage. For each design an expected revenue function is then generated. The designs that maximize the expectation of difference between the revenue and the expenses (payoff) should be pursued. It is common to pursue multiple feasible designs. If R_i is the revenue from the ith design and C_i is the cost of developing and manufacturing it, the expected payoff, P_i, is given by

$$P_i = E[R_i - C_i] \quad (2.1)$$

Since expectation is a linear operator, we can rewrite the above equation as

$$E[P] = E[R_i] - E[C_i] \quad (2.2)$$

If the company utility function* is known, the expected utility criterion can be used as follows:

$$E[U(P_i)] = E[U(R_i - C_i)] = \int\int U(R_i - C_i) f_{R_i,C_i}(r_i, c_i) dr_i\, dc_i \quad (2.3)$$

where $f_{R_i,C_i}(r_i, c_i)$ is the joint distribution of revenue and cost, and the domain of integration includes all regions where the joint probability density function (pdf) has a value greater than zero. Realize that significant uncertainties are present in the early design process, and as a result, simplistic methods such as the one above should be used. Equation 2.3 provides a criterion to rank multiple candidate designs, and the top few can be chosen. It is not necessary to know the exact distributions of R_i and C_i. A well-educated guess with a normality assumption should be enough. A crude utility function may also be preferable to a risk-neutral valuation because uncertainty in the values of revenue and cost is usually large. Performing a rigorous analysis at this stage will defeat the whole purpose of a feasibility analysis. Feasibility studies, in some form or other, should be a constant throughout the design process, as they provide a way of foreseeing difficulties. Expert opinion has a lot of value in performing a simplistic analysis. For example, for each design, an expert can provide an estimate of the probability distribution of the manufacturing cost. If the upper and lower bounds, and the most likely value, are available, one can use a beta distribution, as we will see in Chapter 6.

Another relatively simple but powerful tool in choosing between (or ranking) options under limited information is *dominance*. Clearly, if a design option costs more and generates less revenue than another, it should be

* We will discuss utility functions in detail in Chapter 3. For now, the reader should simply know that utility functions order different alternatives by assigning a numerical value to them that measures their desirability. Utility functions also incorporate the decision-making preferences of the DM under uncertainty.

discarded in favor of the latter. Making this determination is not so straightforward when uncertainty is present. In such cases, one looks at the cumulative density function (CDF) of the payoffs (difference between the revenue and cost) of the design options. Design A is preferred to design B if there is a payoff value, x, where the CDF of P_A is $F_{P_A}(x) = 0$, while the CDF of P_B is $F_{P_B}(x) = 1$. Since a cumulative density is a nondecreasing function, design A's payoff, even in the worst case, is greater than design B's payoff in the best case. This situation is referred to as deterministic dominance. The more commonly observed stochastic dominance occurs when the dominating design's CDF value is lesser than that of the other throughout the domain of payoffs, i.e., $F_{P_A}(x) \leq F_{P_B}(x)\ \forall x$. In cases where dominance can be established, it is relatively straightforward to make a decision. One can simply choose the dominating design without actually finding any expectations, as in Equation 2.3. In Chapter 3, we will discuss in detail deterministic and stochastic dominance in decision situations.

Once feasibility heuristics are out of the way, we can talk about actual design of the product. Realize that feasibility studies may produce multiple candidate designs; they can all be explored in more detail before one is chosen for manufacture. For the rest of the section we will cover the design process for one design. When a design is found to be dominated by others, it can be discarded from the set of candidate designs, and only the remaining ones considered for further analysis. This underscores the importance of constant feasibility studies during the engineering design process. An error caught later in the design process is much harder to fix than one identified earlier, when fewer resources have been committed.

An engineering product design problem is generally too complex to analyze in its totality. It is therefore decomposed into its constituent *design subproblems* (DSPs). A functional diagram of the product helps visualize the product components and the interrelationships among them. It depicts flow of energy, material, and information among different constituents (roughly components) of the product, each of which essentially poses a DSP (Figure 2.1 shows an example functional diagram of a stapler). Chapter 7 describes functional diagrams in a little more detail. The design problem decomposition can be along the lines of the following:

1. **Discipline:** Decomposition along disciplines such as mechanical, electrical, or software, has the advantage that tasks can be easily assigned to the subject matter expert(s). The disadvantage is that a product may or may not be amenable to such decomposition. For example, a motorized robotic arm has both electrical and mechanical components that need to work together. If two separate engineering teams were working on the same robotic arm, conflicts would be possible.
2. **Product modules:** Decomposition along product modules actively considers the architecture of the product in question. It is generally

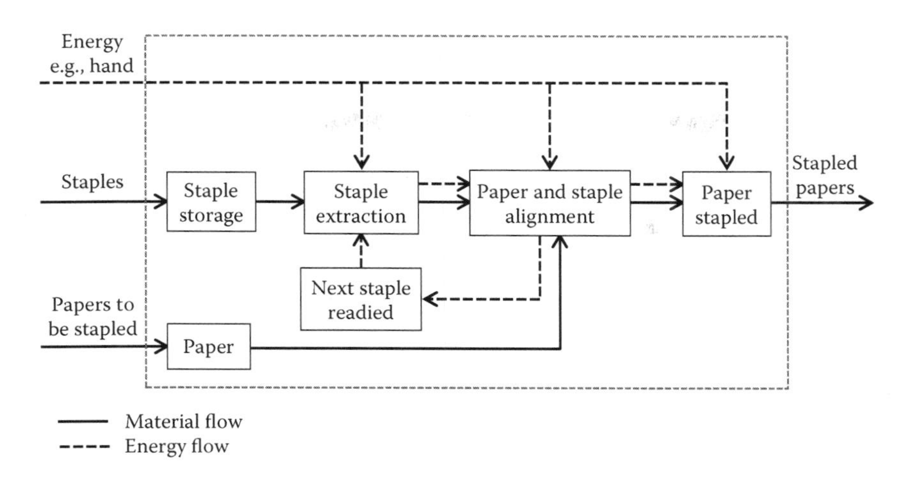

FIGURE 2.1
An example of a functional diagram for a simple stapler.

complementary to discipline-based decomposition. In large design teams one might first decompose along product modules and within the module along disciplines. Such decomposition is also amenable to tracking and scheduling.

3. **Design teams:** In some smaller companies with amorphous structure, different design teams look at different DSPs regardless of the technical nature of the problem. An example is when making one-off products such as semiconductor processing laboratory equipment. Team members need to be trained in cross-disciplinary skills for this decomposition to work.

Most companies implement a combination of the above decomposition strategies, depending on the design problem at hand and the expertise of the engineers and other team members. Significant effort should be spent in making the DSPs as concrete as possible with their interrelationships properly understood. The granularity, how far one should go in decomposition, is determined by the level of clarity about the problem statements of each DSP. For example, the DSP "design an automatic transmission with a given speed ratio, maximum torque/speed rating, and reliability" is too coarse. Similarly, designing just a gear of the epicyclic gear train within the transmission will be too simplistic. A good decomposition would be "design the torque convertor for the given specifications, design the clutches and bands, write software that determines when the computer shifts the gears (based on torque/speed)," and so on.

Most DSPs will have readily available solutions to them. For example, converting rotary motion to linear motion can be accomplished in many different ways; power requirements and geometric constraints can then help quickly narrow them down to the best implementation. Similarly, a power source for an electronic product could be chosen from many already established ones. The solutions, however, still need to work together in a product; details of this interdependence (not necessarily exact implementations) are figured out while drawing the functional diagram and defining DSPs. Realize that the functional diagrams are also subject to updating as the design gets more and more concretized (we will talk about the iterative nature of design later in the chapter). DSPs that have readily available solutions are generally not a priority in the initial engineering design meetings. It is the ones that require development of a new technology or coming up with ingenious solutions that are mostly discussed. Formal engineering design methodology provides the most value for these DSPs.

The conceptual design process is undertaken for each DSP by a team of designers. Ideas are brainstormed, with team members proposing and exploring solutions to them. In a well-conducted brainstorming session each member is able to contribute. Experienced engineers are able to draw upon their experience and recognize similarity with something they have seen in the past. Engineers that are relatively new to the product type are able to propose new "outside the box" solutions. Brainstorming requires every team member to have an open mind—ideas that seem absurd at first can be made to work when others build upon them, and similarly seemingly good ideas can have fatal deficiencies. It is always a good idea to have rough specifications, drawings, and material choices available for each concept proposed.

Engineering design is very much a cognitive process, deeply influenced by biases and predispositions. Coming up with solutions to design problems that are based purely on engineering rationale therefore becomes difficult sometimes. A deeply related concept of how cognition helps propose design solutions is creativity, which is the ability to come up with novel design solutions. A good brainstorming session allows the team members to be creative while keeping engineering constraints in mind. The biases and predispositions also do not always make for an inferior design; when channeled properly, they make a design unique and resistant to errors. Brainstorming solutions to DSPs requires that creativity be encouraged. Also, creativity is not all innate, as it has been traditionally thought to be, but can be learned or stimulated by creating abstractions of an engineering idea. This is because abstractions help get a higher-level feel of the problem at hand. For example, "design a seven-speed automatic transmission" is too specific, while a problem statement such as "design a component that allows changing the torque/speed output of the prime mover" can lead to many creative solutions, for example, the continuously variable transmission (CVT) or an all-electric transmission. Every transformation of an abstract idea to a concrete one and back generates information that can be utilized in design. One starts recognizing patterns in the repetitive features that can be made part of the formal design. Other methods to stimulate creativity are to make associations between different concepts and to ask many questions related to the product under question until key engineering issues are identified. We will return to this topic of brainstorming and creativity in Chapter 7.

Concept generation cannot always be fully accomplished in one meeting. Many times the team members need to go back and think independently about the design before another session is conducted. In such cases the importance of information collection from external sources cannot be overemphasized. Rarely does a firm encounter a design situation that requires coming up with fundamentally new concepts. Designers should be willing to look into prevalent literature in the field and talk to experts. Attending conferences and talking to consultants and other designers is a critical part of creating and maintaining a knowledge base. Research papers and patents prove to be valuable sources of information. They also help realize where the state of the art lies and if intellectual property needs to be licensed from someone who has already solved the problem. It is possible that the solution to a particular problem is not found through information collection; however, solutions to closely related concepts can sometimes prove to be helpful.

In today's highly specialized environment, once the design is agreed upon, the detailed design is generally attacked by a specialist in the field. He or she will create what is called a detailed design for the DSP with all the specifications, such as material choice and dimensions. For example, in mechanical engineering–related problems this step will entail making a CAD drawing. This allocation of labor for each DSP also increases the responsibility of the specialists, since the amount and frequency of information sharing should be carefully considered. Specialists must be able and willing to collaborate when working on closely related DSPs. While initial design parameters are mutually arrived at in a concept generation meeting, any change in the design or its implementation should be communicated to the rest of the designers. Collaborating designers must also be respectful of each other's ideas and suggestions. To that effect, most companies entrust a project manager with the responsibility of making sure that the group members understand their duties and meet their collective requirements. The designers eventually look to combine the solutions they have found for the DSPs into a concept of the product. As the design process progresses, the product concept gets significantly more concretized and starts looking more and more like a design. Proper documentation of the design process ensures that the information gained is turned into knowledge for the next cycle of design.

A global perspective is very critical in highly interdisciplinary system design problems of today. Engineering design must address global issues of customer requirements from the product and the overarching metric of profit making for the company. The final design of a product determines what processes will be involved in its realization. Most designs are tested by building prototypes before large amounts of resources are committed to their mass production. The prototype is then tested for its ability to perform intended functions in actual use. Often errors are found in design or manufacturing that must be rectified. Reliability engineering concepts of infant mortality and accelerated testing are very relevant here, and these will be covered in Chapter 4. Only after the deficiencies in prototype are fixed and the workers are trained does a product go into mass production.

Since all decisions made by humans involve uncertainty and are inherently constrained by the inability of humans to consistently make optimal decisions, designers should always account for them. Proper training and experience in probabilistic methods is very helpful. Designers should also strive for continuous improvement in their knowledge level and skills. Not just the product itself, but its design process also should be such that human errors are minimized. This can be done by accounting for foreseeable and unforeseeable errors introduced by human operations. The uncontrollable uncertainties, whether due to human error or environmental factors, should be taken into account as well. It is no coincidence that when assessing utility functions, tremendous care is taken to account for preferences under uncertainty.

As a final note, a structured approach to design provides many advantages. Sometimes there is resistance in industry, with designers complaining that they do not have the time to go through all the steps in formal engineering design. Pahl and Beitz (1995) argue that a systematic approach to engineering design is necessary because these steps are taken inadvertently anyway. A systematic approach ensures that nothing of importance is overlooked. A lot of times inefficiencies and rework in companies are a result of bad practices that have accumulated over the years. Engineers and managers involved have found a way to reconcile with these bad practices and become defensive of these methods. A systematic approach is a great way to break away from that tradition and return to profitability by realizing efficiencies in overlooked areas. Finally, a systematic step-by-step approach also helps with project planning and scheduling, allowing for further improvements.

2.2.1 Fermi Method

Magrab (1997) discusses an insightful approach called the Fermi method, which is worth mentioning here. He gives the example of physicist Enrico Fermi, who was known to be able to arrive at good estimates of numbers and parameters from very limited information. The most well-known example is when he was able to estimate the power of the first experimental atomic bomb by observing how far bits of paper were dispersed by the shock wave from the explosion. Magrab argues that Fermi was good at providing such estimates because he was able to combine new information with his prior knowledge and use logical deductions. Many times estimates of mathematical quantities are required in engineering design with little or no information. If reasonable assumptions are made over what is already known, a ballpark value can be reached because the errors in assumptions tend to cancel out.

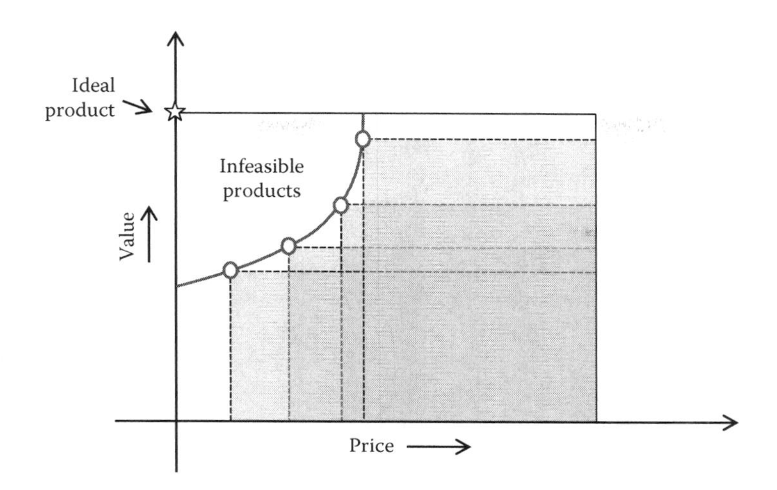

FIGURE 2.2
A simplistic representation of how multiple products can capture most of the demand.

2.2.2 Product Platform and Product Placement

A product platform is the basic essentials of a product from which a set of products with slightly different attributes can be derived. Multiple products have the capability of capturing a large portion of the market as compared with a single product. The products achieve different levels of attributes, starting from the base product platform; by varying the design parameters just enough, price is also varied accordingly. Figure 2.2 shows a simplistic representation of how this plays out in reality. Customers trade off price with the value a product provides. Clearly a product that provides maximum possible value at the minimum possible price will capture all the market (shown as a star in the figure), but such products are more often than not infeasible. It is not possible to manufacture them because of technical or financial infeasibility, or both. Unavoidable trade-offs therefore must be made, which results in a Pareto front,* shown as the curved line in the figure. It is impossible to manufacture a product that can *simultaneously* improve value and reduce price for a product on the Pareto front. Each product is shown to have a rectangular region to the right of it. All customers who are willing to buy something with lesser value *and* higher price than a product (shown on the Pareto front with a circle) are included in the rectangle associated with that product. As more products are added on the Pareto front, the number of customers covered also increases. Indeed, there is overlap (same customer can purchase different products), but each product adds unique customers to

* We will encounter Pareto fronts many times in this book. A curious reader may want to peruse Chapter 3 for a better idea of what they are and how they are relevant to engineering design.

the demand. This, however, cannot be guaranteed if dominated products are offered (imagine a product to the right and below a point on the Pareto front).

The decision of where to offer products on a Pareto front is not an easy one. The process is called *product placement*. In addition to the portfolio of products offered by the manufacturer itself, one has to consider competitors' products. Product portfolio planning also incorporates design and manufacturing considerations. An 80-20 rule is often cited; that is, roughly 80% of the market coverage is realized by 20% of the products. Therefore, to justify adding new products to the portfolio that may cover only a small incremental market share, one has to carefully consider all the ramifications. Automobile manufacturers, for example, recover a large portion of their profits from the sale of high-end luxury cars. This is because even though they form a small segment, the profit margins are extremely high.

To have a perceptible distinction in products, they must have differentiating features. Obviously this comes at reduced economies of scale. The differentiating features therefore must not be hard to implement or to maintain to improve profitability significantly. Good product platform thinking actively considers this problem. All products share the basic design and many components, which form the essentials of the product, that is, the product platform. The differentiating components are such that they can be easily integrated into the products. Some tablet PCs manufactured by the same company, for example, share the same form, screen, and processor but differ in storage capacity or availability of a mobile network antenna. As a result, products can be realized that are not that different to design or manufacture, but are perceived as being different by the customers.

Product platform design also has a temporal aspect. A well-planned product platform is flexible enough to allow for evolution of products. As customer needs change with time, the product attributes evolve together, hopefully maintaining or improving on the market coverage that previously existed. Customer preferences in some attributes are monotonic in time, which allows for slow permeation of technology from a higher-end model to a lower-end model. For example, antilock braking systems (ABSs) were restricted to luxury cars until the late 1990s. Since the technology is now mature, mass production is cost-effective and can be introduced in lower-end cars. Since ABS has become almost standard on all cars today, it no longer acts as a differentiating factor among products. This evolution was possible because ABSs were designed to be easily integrated into the braking systems of most automobiles.

An eye should also be kept on commonality when designing a product platform. Commonality refers to the extent to which products share components or manufacturing processes. Commonality is a well-studied area of design literature. Modular products lend themselves well to commonality measures, as has been seen in many design situations. Khajavirad and Michalek (2007a, 2007b) differentiate between commonality studies on two grounds: commonality as a metric to determine (qualitatively) how much similarity exists between products in the same product line and commonality as a quantitative metric to be optimized in a design decision problem. Having commonality as an objective of optimization, however, is always going to be questionable. This is because not only is it hard to come up with a comprehensive metric for commonality, but a manufacturer does not care about commonality in itself; it is the effect on other, more tangible attributes, such as manufacturing costs and profits, that is the main focus. Considerations of commonality are, however, a must in product platform design.

2.2.3 Engineering Design Is Iterative

Very few designs succeed at the very first attempt. Analysis shows that most designs fail to meet the specifications in some way or another due to small errors propagating as the design process progresses. These errors can appear in the implementation of any of the steps in the product design or manufacturing process. With planning, virtually all of these errors can be caught when they happen or shortly afterwards. Expectedly, as time progresses and more and more resources are committed to a product, the cost of rectifying mistakes increases exponentially. This underscores the importance of periodic inspection and oversight, coupled with routine feasibility studies of each decision. Some of the reasons designs fail to meet specifications at the very first go are

1. The product planners did not agree on deliverables; their individual decisions conflicted, leading to a product that does not perform its intended function correctly.
2. Engineering mistakes were made; dimensional, material, or manufacturing specifications were incorrect.
3. Customer requirements changed, which required redesign. If the design caters to a specific order from the customer, most of the time the mistakes can be rectified at the customer's expense.
4. Costs structure changed; materials became less or more expensive.
5. Uncertainty gets resolved unfavorably in terms of legislation, weather, customer demand, mergers, and acquisitions.
6. Manufacturing errors are made/found. While manufacturing mistakes tend not to be as critical as design mistakes, they must be avoided. In the late 20th century, one of the reasons Japanese automotive firms were able to outperform American ones was because they had strict quality controls during manufacturing.
7. There is a difference of opinion on how to evaluate designs. Even a design that achieved all its deliverables may be considered failed because there is no consensus on how to evaluate it.

Since significant savings can be made in time and money by catching errors early, a design process, or even a simple step within it, requires iterations and revisiting. Steps earlier in the design process, such as concept generation, are therefore more critical to the process than later ones, such as manufacturing and assembly methods used. Researchers almost universally agree that most of the cost of a product is fixed in the design stage itself (~80%). The most important reason is that it precedes other steps. Good project managers always revisit steps in product development before moving further.

2.2.4 Mathematics and Design

Mathematical methods in design are pervasive. The ability to represent an engineering system in a mathematical model opens up a lot of avenues for using mathematical tools: optimization, probability theory, simulation, and functional transforms, to name a few. Design of most product elements—from design of a mechanism for a given time ratio to that of an electronic circuit to solve navigational differential equations—is mathematical. It is the direct application of scientific theories (usually mathematical in nature) to real-life problems. In addition to the ability to design product elements, mathematical modeling and manipulation of design variables within the mathematical model also help locate avenues for improvements in a design. For example, models of demand predict what product attributes will be desirable to customers and, hence, profitable to the manufacturer. These attributes cascade into the functional requirements of the product during the concept development phase. The functional requirements lead to detailed design where engineering specifications (design parameters) are determined. If a mathematical representation of how the design parameters affect demand is available, mathematical optimization can find the best values for the design parameters. Quantitative methods therefore aid design decision making immensely. However, they are by no means a replacement for intuition and expert judgment. Noted mathematician John von Neumann once purportedly said: "If people do not believe that mathematics is simple, it is only because they do not realize how complicated life is."

It is not that mathematics loses value in a complicated interdisciplinary design process. The complexities are sometimes too hard to be mathematically representable, and as a result, approximations are made. If these approximations are defensible and the designers are aware of their limitations, they do not pose too much of a problem. Otherwise, catastrophes can result. Some common mathematical simplifications that can have far-reaching (usually negative) implications on design are

1. A function is linear in the design variables.
2. A function is monotonic.
3. There is no stochasticity in design variables.

4. There is no stochasticity introduced when attributes are calculated from design variables.
5. A random variable is normally distributed.
6. Two random variables are independent.
7. Time to failure of a component is Weibull distributed.

2.2.5 Optimization in Engineering Design

Engineering designs need to be optimal or close to optimal. A design endeavor that starts with a customer needs statement leads to many, sometimes infinite, feasible designs. A feasible design is one that satisfies all the requirements (also mathematically referred to as constraints). An optimal design is not only feasible, but it also maximizes a carefully chosen performance function. In the decision based design paradigm this performance function is the utility function of the designer or that of the idealized decision maker representing preferences of the manufacturing firm in the enterprise context. An optimal decision can rarely be chosen from all the feasible designs by simple analysis, except for some very simple problems. Optimal designs are found by sophisticated optimization algorithms that manipulate the design variables until a solution is found that cannot be improved further. Optimization methods are continually being improved and refined in the kinds of problems they can tackle, how fast they can converge, and the quality of solutions they find. The generic mathematical formulation of an optimization problem is

$$\begin{aligned} \text{Minimize} \quad & f(\mathbf{x}) \\ \text{Subject to} \quad & \mathbf{x} \in X \end{aligned} \tag{2.4}$$

where f is the objective function that is to be minimized, $\mathbf{x}$ is the decision or design vector, and X is the feasible set. X is usually an intersection of sets defined by many equality and inequality constraints. Optimal design formulations should also consider what algorithm will be best suited to solving a problem. The properties satisfied or not satisfied by the objective function and constraints such as convexity and differentiability allow use of certain techniques while precluding others. Constraints should also be well defined. A designer should understand the physical meaning of each constraint and if it can be relaxed. In a highly constrained problem, even finding a feasible solution is very difficult. Sometimes choice of algorithms also restricts how a design problem is formulated. We will cover design optimization in detail in Chapter 5.

Mathematical optimization can range from being extremely useful to extremely impractical, depending on the stage that the engineering design is in. A limitation of optimization in engineering design is that it requires a very good mathematical model of the engineering system. These models are usually very difficult to create, and the effort required may negate the benefits. In our opinion, this limitation must be kept in mind before optimization is applied in decision making. As a general rule, optimization is beneficial in the enterprise-level decision-making stage where high-level objectives are optimized subject to simple constraints. It has limited applicability in the concept generation phase since creativity is the critical element there. It has the ability to complement concept selection by human designers but cannot replace them. Mathematical optimization again becomes indispensable in detailed design where crisply defined DSPs need to be solved in an optimal way. Mathematical models are generally available; for example, deformation of a structural element to an applied force can be mathematically modeled relatively easily.

Some commonly encountered problems that can reduce the applicability of mathematical optimization to engineering design are badly modeled constraints, lack of knowledge about the input–output relationship between design parameters and objectives, incompatibilities between different DSPs that are not modeled, and lack of useful properties in the model (e.g., differentiability).

2.2.5.1 Optimal Design Formulations

Design optimization problems can be formulated in basically two ways: all-in-one (AIO) formulations and target-based decomposed formulations. AIO formulation refers to designing a product (i.e., the mathematical counterpart of it) in its entirety. All-in-one design formulations, when implemented correctly, have the advantage that the interdependence between different DSPs is explicitly considered. The obvious trade-off is that a comprehensive mathematical model of the design problem must be available.

In target-based design optimization formulation, design is usually segregated into system-level and subsystem-level design problems, as in analytical target cascading (Kim, 2001). System-level attributes (e.g., firm profit, market coverage) are optimized with minimal information about design constraints. These optimal attribute values are then passed to lower-level subsystems as targets. The lower-level subsystems try to meet these targets by minimizing the error between the best possible attribute levels they can achieve by changing the design variables and the target prescribed to them (Figure 2.3). As one goes further down in the hierarchy, subsystem-level problems get closer and closer to basic engineering problems (dimensions, materials, etc.). Different subsystems also share common design variables that must be matched for consistency. For example, if one subsystem problem involves piston design while another designs the cylinder block, the relevant dimensions must match; otherwise, the engine cannot be assembled.

Target-based optimization formulation is, by definition, iterative. At every iteration, targets are passed down and variable values are passed up. Higher

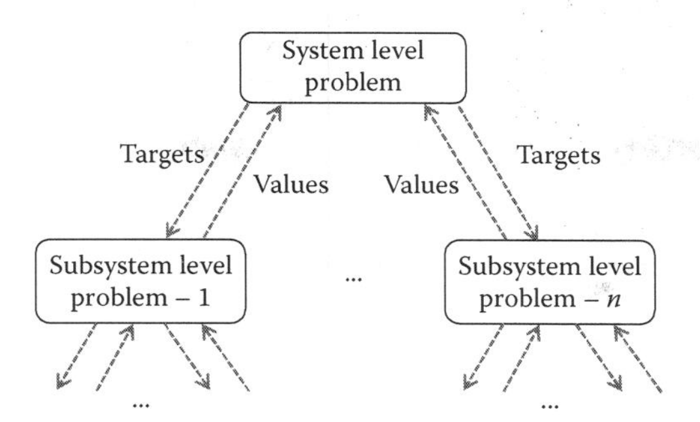

FIGURE 2.3
Multilevel target-based design. Higher-level problems pass targets to lower-level subsystem design problems, while lower-level problems return variable values.

levels have to not only prescribe attribute targets, but also collect inputs from different subproblems and set consistency targets between conflicting values of the same variable that two or more subsystems share. Different implementations use different approaches for target setting as well as maintaining consistency. Target-based multilevel design optimization has a natural counterpart in actual product design in firms. Most of the design projects are aimed at meeting a target specification in a product or its components. These targets are cascaded down to individual engineering decisions. In a hierarchical product breakdown, target-based design is evidenced more and more as one gets further down the design chain. Since the arrow of causality is an important consideration in design, target-based design optimization helps represent it in a mathematical way. It is important to know, however, that the decomposed implementation in target-based design optimization may not always converge to an optimum or even a feasible system. Many times the computational effort required in coordination far exceeds the gains from the decomposed nature of the problem.

2.2.6 What Drives Engineering Design Research?

Engineering design research is driven by the need for it—both in industry where it is applied and in academia where fundamental problems in engineering design research are solved. Although the main objective in industry is profit, many times, long-term research is undertaken without clear profit implications, in the research and development (R&D) departments of many companies. By investing in R&D, a company tries to stay competitive in the market by developing new technologies or creating barriers to entry by patenting intellectual property. Independent research in such cases is usually encouraged with little supervision. Other major settings for engineering design research are government labs and universities where sponsored projects are undertaken to reach a fundamental understanding of engineering design. The objective is not necessarily to make money, but to generate a body of knowledge that can help the industry (with ideas and trained individuals) and consequently a nation's economy. Engineering design research helps enable efficient engineering design practices built upon observing what is currently being done, learning from it, recognizing patterns—both positive and negative—and proposing solutions. It focuses on details of a real-life engineering design situation and proposes solutions to these through analytical means, which may entail qualitative reasoning or mathematical methods. Engineering design research is therefore a service offered to the industry by researchers in a university or a government lab. Various fundamental solutions to problems faced in engineering design have traditionally come out of research institutions.

Whether or not engineering design is a science is open to interpretation. For the most part, engineering design is not exactly repeatable and reproducible, which any scientific theory has to be. However, as we saw in the previous sections, engineering design does follow a structured approach and should be studied and implemented as such. To identify research problems in engineering design, one must understand the information flow within an organization engaged in it. This level of understanding of information flow helps an engineering design researcher understand the role of decision makers and how to best perform market surveys, devise new methods for demand modeling, propose solutions to organizational planning (for example, the design structure matrix, which we will study in Chapter 7), understand logistical issues, and most importantly, identify and propose solutions to technical engineering challenges.

2.2.7 Identifying the Scope of Design

No design can be expected to perform all the functions imaginable. One does not expect a laptop computer to be able to peel a potato! Fortunately, customers and end users generally do not have such expectations. Many design requirements tend to conflict with each other and make the job of the designer harder. A design that has a relatively narrow focus is amenable to mass production and helps realize economies of scale. It allows designers to find solutions to design problems efficiently. Later in this book, when we get into the mathematics of utility theory, reliability, and optimization, we will see that defining the correct scope of design is not only desirable, but imperative.

A well-defined scope helps with the mathematical formulations as well as optimization. Utility functions are also best assessed when the range of negotiability of the attributes (best and worst levels) is well defined. In optimization, various convexity and monotonicity properties that aid optimization algorithms depend on a well-defined constraint set. Functions are required to satisfy these properties within the feasible set only. A function is more likely to be convex, for example, in a small local region than over a large region. When using heuristic techniques, a smaller feasible set is even more critical for finding optimal solutions.

There are some well-known exceptions, though. The multitool Swiss army knife is an example of a product with many functionalities. The essential thing to note is that the different functionalities of the knife do not conflict. The different elements can all be considered tools, and they do not have very different operating requirements. Adding many functionalities adds new constraints to the design problem, and one tends to lose optimality in one functionality. For example, a large knife will outperform a Swiss army knife at cutting harder material of a large size. A heavy-duty screwdriver can be used on screws requiring high torque to tighten or loosen. Flexible design, of which the Swiss army knife is an example, however, is also an ongoing area of research, where a product is designed to work in different modes and configurations. The compromise in performance in each mode is also well documented.

2.2.8 Fundamental Objectives vs. Means Objectives

Every problem or subproblem of design has some fundamental objectives that must be achieved. *Fundamental* objectives are relatively few in number, but they are what really concerns a decision maker. *Means* objectives act as stepping stones to the fundamental objectives. Consider the design of a car. The fundamental objectives a customer generally wants a car to satisfy are safe transportation at a reasonable price. There are three objectives built into that statement: transportation, safety, and price. The car must provide transportation, the car must be safe, and the car must not cost more money than the customer is willing to spend. The means to make a car able to move is to incorporate a prime mover (usually an engine). Safety features reduce the likelihood of physical injury by avoiding an accident or by reducing the severity of injury if an accident does happen. Finally, price is a function of other attributes, and the manufacturer can fine-tune it to control demand.

Traditionally, the prime mover in a car has been an internal combustion engine; safety features include airbags, seat belts, ABS, etc.; and price is a function of all the design and manufacturing expenditures plus a profit margin. A company can, however, choose to make a fully electric vehicle and still meet the fundamental objective of transportation. Safety features can include basic features such as airbags or even sophisticated collision avoidance systems. There are many ways to address the fundamental objective of reduced fuel consumption as well. However, when one starts looking at implementations of designs that address these fundamental objectives, it is easy to lose focus and get anchored to a particular implementation. Design solutions that satisfy means objectives should always be considered expendable in favor of solutions that directly attack the fundamental objectives. Having said that, the role of means objectives is by no means trivial. Once a system structure is determined and different decision makers have been entrusted with the job, means objectives start looking more and more like fundamental objectives. Sometimes in a decomposed problem, a designer may not even be aware of the main objectives. Many researchers are active in this area of design problem decomposition so that the advantages of focused decompositions are not affected negatively by the lack of understanding of fundamental objectives.

2.2.9 Decision Makers and Consensus

In the final decomposed form of a design, the decision makers should all be known, and the decision makers themselves should know the scope of their decisions. They should be aware of what performance function they are working for, what the time constraints are, and what resources they have available at their disposal. One of the challenges of involving multiple decision makers in a team is to achieve equity. Equity refers to the state where all the concerned decision makers are equally happy. In a design situation, considering the isolated view of a firm, equity will refer to all the concerned individuals being equally content with the design. When we zoom out a little bit and include the customers as well, equity will mean that the firm and the customers are equally happy. Since we have a normative way of measuring an individual's happiness, the utility function, it seems initially that we should be able to achieve it. Achieving equity in decision making is not only a challenging task, but it cannot be done unless interpersonal comparisons of preferences are made.

In social science, interpersonal comparisons of preferences are known to be inherently flawed. Arrow's impossibility theorem (AIT; Arrow, 1950) states that no voting system can satisfy some basic commonsense axioms. Kirkwood (1979) even showed that attempts to address equity actually take one away from a global optimum. It is important here to mention that AIT does not apply readily to engineering design. Scott and Antonsson (1999) provide compelling arguments as to why this is the case. They argue that an engineering firm can in fact represent an idealized decision maker; therefore, the issue of addressing multiple decision makers can, at least in theory, be avoided. Furthermore, in social science, sovereignty of all decision makers is assumed. There is no reason to assume this in engineering design because design team members have different backgrounds, seniority, as well as experience levels. Scott and Antonsson do caution that the issue of interpersonal comparisons does play an important role, even though it does not preclude optimal design.

Another work showing the way to reconcile with AIT involves the use of cardinal utilities. Keeney (1976) showed that it is possible to satisfy the commonsense axioms that Arrow set out to satisfy, with cardinal utilities. Cardinal utilities measure strengths of preferences in addition to providing ordinal rankings. Keeney's method does not avoid interpersonal comparison of preferences but provides a way to systematically consider everyone's preference. He also showed that the only way to combine the utilities of the decision makers is to use a linear combination of them in making decisions. The relative weights given to each decision maker are decided by consensus or by a "benevolent dictator." Since the firm can be considered an idealized decision maker, finding a group utility function is very much possible.

The decision makers must decide, for each decision, whether to maximize their personal utilities or the firm's utilities. This depends entirely on the scope of the decision. If a decision affects the whole firm and its finances, then the firm's utility function must be used. For design decisions with limited scope, personal utility functions will give the better results.

2.2.10 Some Thoughts on Manufacturing

While the focus of this book is on design, manufacturing cannot be ignored. Design of a product, for the most part, determines what the manufacturing methods will be. In fact, design specifications ask for such information. Designers therefore must consider the cost and ease of manufacture when selecting from different alternatives. The formal study is termed *design for manufacture*. Manufacturing is the process that concretizes design into a physical artifact and is therefore very labor-intensive. While most of the expenditure and effort happens during manufacturing, most of the manufacturing costs are *determined* during the design phase itself. Designs must consider what manufacturing methods will be used and their tentative cost. Design for X techniques identifies different implications of a design and focuses on them. For example, design for assembly strives to consider assemblability of the components while still in the design phase. Many designs that seem good on paper are hard to manufacture or assemble and must be redone.

2.3 Uncertainty in Design

Uncertainty affects almost all our decisions, and engineering design is no exception. Design under uncertainty poses two major problems: (1) Only the statistics of the uncertain design variables can be controlled, and (2) the outcomes of design decisions (outcomes resulting from the choice of statistics of the design variables) cannot be predicted perfectly. Needless to say, uncertainty needs to be mitigated or, when that is not possible, reconciled with. Consider the design of a pin that goes into a joint in a slider-crank mechanism, as shown in Figure 2.4. The design calls for specifying the cross-sectional radius and the material that will handle a certain amount of shear stress. Since there is always manufacturing uncertainty, there is no way to know what radius to specify to achieve a particular shear strength. Since material properties are also uncertain, the uncertainty in the shear strength

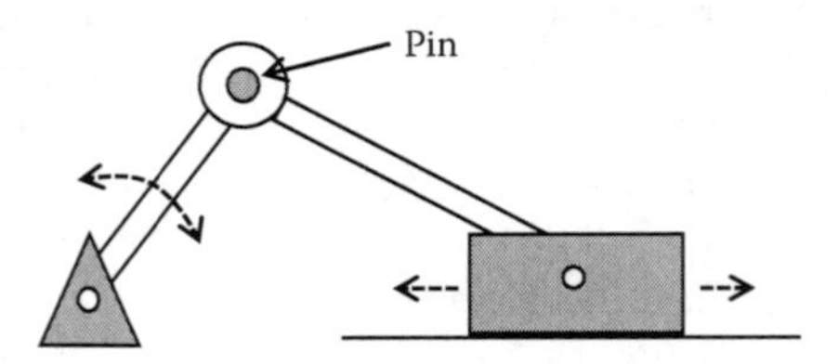

FIGURE 2.4
A slider–crank mechanism.

of the pin increases further. Operating conditions also make the stresses in the pin stochastic, adding uncertainty in successful operation of the mechanism. A common technique in such cases is to include a factor of safety, that is, specifying a bigger dimension than necessary. Doing that, however, will increase the weight of the pin, increase the cost, and may even cause failure through some other mode, for example, increased stress on the motor. A design problem involving uncertainty can be analyzed normatively using decision analysis. In the next chapter we will learn methods to make such decisions using a carefully assessed utility function. For example, if U is the utility function over stress, s, and μ_{X_i} is the mean diameter (X) of the pin for design i, then the mean diameter to be chosen (for a given material) that maximizes expected utility from the design will be given by

$$\mu_X = \arg\max_i \int_0^\infty U(s(X)) f_X(x \mid \mu_X = \mu_{X_i}) dx \tag{2.5}$$

Notice that the pdf of the stress changes depending on the mean chosen, as expected.

Uncertainty is classified into two kinds: reducible (epistemic) and irreducible (aleatory). Reducible uncertainty can be reduced by performing tests or collecting more information; irreducible uncertainty cannot be. In the above example, the uncertainty regarding the material properties can be reduced by performing strength tests on specimens from the same materials supplier. Value of information studies can also be performed to see if performing the tests is worth the effort and money. The uncertainty because of the manufacturing uncertainty, however, cannot generally be reduced. Buying precision machines that reduce this uncertainty is justified only for higher-end products with bigger profit margins. Clearly, uncertainty cannot be completely eliminated. As a result, the design decision must be made to reconcile with these uncertainties. As we will learn in the next chapter, a decision that maximizes the expected utility should be made. For the above example, we would first quantify the manufacturing uncertainty, for example, specify a statistic of the diameter (e.g., its mean) that gives the highest expected utility from the mechanism given the resulting distribution of its strength. This can be repeated for different materials while also incorporating the uncertainty in their strength.

Identifying sources of uncertainty is not always an easy task. While tolerances are known in machining operations and their effects easily modelable, extraneous factors such as weather conditions, worker strike, and changing decision maker preferences are sometimes not explicitly considered. Incorporating every possible source of uncertainty will inevitably increase the time and effort required. A meta-decision can be made to first choose which sources of uncertainty to consider and which to ignore. A simple rule of thumb is to approximate the impact of a given source of uncertainty on the variance of the performance function. Tornado diagrams are a very powerful tool to present this information in a graphical form. The reader is referred to Clemen (1997) for a treatment of tornado diagrams. Once the sources of uncertainty are identified, they are modeled using probability distributions.

Uncertainty also plays a part after a product is manufactured and put to use. Promised attribute levels cannot always be met far into the useful life of the product. Furthermore, products also fail because of operator error or suboptimal operating conditions. Warranty burdens ensure that most manufacturers consider reliability of products carefully. Predicting reliability requires probabilistic modeling of operating conditions and the likelihood the product will be able to successfully operate under those conditions. Design for reliability is a field that actively considers the reliability of the product manufactured. The cost of additional reliability is traded off with the cost of repair or replacement if the product fails. We will consider reliability engineering principles in detail in Chapter 4.

2.3.1 Notation

For the purposes of this book, we will assume that the reader is comfortable with basic concepts in statistics. Here we provide a brief background to refresh the reader. A random variable is represented with a capitalized letter such as X, while its realizations are represented using a lowercase letter (x). The probability density function (pdf) is a function that determines the likelihood of the random variable lying between two points in its domain (which is also referred to as the support of the distribution). For example, the probability that X lies between two points, a and b, on the real line, is given by

$$P(a \le x \le b) = \int_a^b f_X(x)dx \tag{2.6}$$

where $f_X(x)$ is the pdf of X. A cumulative density function (CDF) at a point x is the probability that the random variable assumes a value less than or equal to x. A CDF can be completely determined by the pdf of the random variable.

$$F_X(x) = P(X \leq x) = \int_{-\infty}^{x} f_X(t)dt \quad (2.7)$$

In multiple dimensions we look at joint CDFs of multiple variables. A joint CDF is a two-increasing function of the random variables. Joint CDFs can be represented using copulas by using the marginal distributions of the constituent variables. These concepts are presented at various points in this book wherever relevant, particularly in Chapter 6.

Problems and Exercises

1. What is engineering design?
2. Who is central to engineering design? When did this change in outlook toward design take place?
3. What is the role of decision analysis in design?
4. Identify the role of feasibility analysis in early stages of design.
5. While performing initial feasibility analysis, it was assessed that two designs, A and B, have normal payoff distributions with {mean, standard deviation} of {4 million, 1 million} and {5 million, 2 million} respectively. Which design should be selected using a risk-neutral valuation? Which design should be selected if the standard deviation for design B is 4 million? What does this tell you about risk-neutral valuations vs. using utility functions?
6. How does uncertainty affect engineering design?
7. What is a design subproblem (DSP)? Why is it important to decompose product design into DSPs?
8. What are the three common ways to decompose product design into DSPs?
9. Comment on how creativity affects concept generation and selection.
10. Comment on the value of iterations in the design process.
11. What is product platform thinking? How is it important for the long-term success of a product?

12. What is commonality in engineering design? How should it be approached? What are the pitfalls of making commonality an objective of optimization?
13. What is Arrow's impossibility theorem? Discuss its relevance (or irrelevance) to engineering design?
14. What are some commonly used mathematical simplifications that plague engineering design?
15. Discuss the role of optimization in engineering design. Why is it helpful? Also discuss cases where the efforts required in mathematical modeling for optimization may not be justified.

3

Decision Analysis with Multiple Attributes

3.1 Decision Making under Uncertainty

All engineering design processes can be looked at as a series of decisions made by the stakeholders involved. For example, consider mechanical engineers looking to improve an existing automatic transmission. Mechanical engineers can decide to use a new material for gears, use a different manufacturing process, update control software, or even pursue a radically new design—in hopes of making it more reliable, cheaper, or both. They may also choose to do nothing to the existing design and move to another product altogether. These decisions need to be evaluated based on their impact on the attributes with which the DM is concerned. In a for-profit firm, the overarching attribute is usually profit, but surrogates of reliability are also sometimes used. It is clear that for sufficiently complicated products, the number of decisions that can be made are enormous, possibly infinite. Consequently, it is not always possible for designers to *actively* make each decision themselves. In helping the DM make decisions, we need a formal understanding of his or her preferences. The benefit of a formal understanding of a decision maker's preferences is the implementation of these preferences to make decisions by the decision maker himself or herself or by a facilitator.* One can also embed his or her preferences in a mathematical function so that the process of decision making can even be automated on the computer. The decisions are still being made, but by a different, possibly virtual, entity that (hopefully) mimics the actions of the decision maker. Since an error in modeling the preferences of the decision maker can lead to a grossly incorrect decision, a lot of emphasis is placed in decision analysis on finding the functional forms that truly model decision maker preferences. In this chapter we will study in detail how all this can be accomplished and about what pitfalls we should be careful.

Formal study of decision making under uncertainty has been around for a long time. Originally, it was mostly studied in the context of statistics; that is,

* A facilitator is an expert in decision analysis who elicits the decision maker's preferences and helps him or her make decisions consistent with these preferences.

a good decision statistically gave good outcomes. This worldview changed with the resurgence of Bayesian statistics during the mid-1900s, where the decision maker and his or her degree of belief about the likelihood of an event (probability) came to the forefront. The term *decision analysis* was coined by Ronald Howard (1966). It is termed a normative field in that it does not describe what a decision maker would do, but rather what the decision maker should do. This is fundamental to the understanding of the distinction between a good decision and a bad decision. A field that claims to be normative must be able to help us make *good* decisions. In the following we first define what a good decision is. Then we discuss the ground rules that underpin decision analysis, with particular emphasis on the role of uncertainty. We will show that uncertainty does not really affect our ability to make a good decision.

3.2 Good Decisions and the Role of Uncertainty

Decisions almost always involve uncertainty, which affects outcomes. Uncertainty in the nonmathematical sense can follow any of the various definitions. It could mean that we do not have complete knowledge of possible outcomes, or it could mean that we do not know how likely they are to be realized, or both. Uncertainty can also be construed as the case where the decision maker does not know what he or she wants. Uncertainty can be classified into aleatory (uncontrollable, irreducible) and epistemic (controllable, reducible) uncertainty. Aleatory uncertainty or irreducible uncertainty is the inherent randomness in a system and cannot be reduced, the classic example being the flip of a coin or the temperature during a certain time of the day 5 years from now. Epistemic uncertainty is the uncertainty present because of lack of knowledge and can be reduced by collecting more information; for example, the uncertainty about the lift generated by an airfoil can be reduced by running computational fluid dynamics (CFD) codes or by performing wind tunnel experiments on a model. Reducing epistemic uncertainty always costs resources, which is why we sometimes perform value of information studies to see if it is justified to collect more information.

The immediate consequence of the presence of uncertainty in a decision scenario is that even the best decision leads to a multitude of outcomes, some of which could be bad. How do we make good decisions then? Turns out that we first need to define what a good decision is, under uncertainty:

> **Good decision:** A good decision is the one that maximizes the probability of a good outcome.

Notice that the above definition talks about the *probability* of a good outcome, not a guaranteed good outcome. One crucial thing one must be careful of while evaluating a decision is that outcomes generally do not say much about the quality of the decision. This is because a good decision can have a bad outcome, and a bad decision can have a good outcome. After all, that was something we set about to reconcile with anyway! Everyone can recall an incident from his or her life where he or she made a good decision that resulted in a bad outcome. Consider your friend Tammy getting new tires on her car after finding out about impending bad weather. She can, however, still get into an accident if the tires are not installed correctly. Similarly, she could also continue driving the automobile with bad tires and end up not having any accidents because she is lucky. Clearly, getting new tires is a good decision, if she can afford them, but does not guarantee a good outcome. In engineering, decisions are multilayered and complex, and involve interaction of human and machine elements. As a result, it is often not clear right from the onset if a decision is good or bad. Therefore, a decision analyst should understand the unavoidable risks encountered even when good decisions are made. More importantly, he or she should not be overwhelmed by the many decision variables, attributes, and sources of uncertainty but try to make a structured decision founded on a defensible set of rules. Decision analysis helps us do just that.

Recall that there is an emphasis placed on the term *normative* in decision analysis. It qualifies the field as being normative or prescriptive, as in what a decision maker should do, as opposed to what a decision maker would do. This distinction is important because being able to make a rational decision has little to do with our instincts, which most people use to make everyday decisions. For everyday decisions it is not critical to be able to make the best possible decision; it may not even be advisable to pursue it, considering that the effort may not be justified. However, when decisions involve extreme outcomes, such as possible loss or gain of large amounts of money, loss of life, and impact on many people, we need methods that can help us make good defensible decisions under uncertainty.

Uncertainty in engineering and sciences has a specific meaning. It refers to a physically discernible mathematical quantity that does not always take a fixed value, but one from a range of values called realizations. We generally do not talk about what the weather is going to be like tomorrow; instead, ask "What will be the temperature in degrees Celsius at 3 p.m.?" or "What will be the relative humidity at noon?" This helps in clearly defining what aspect of the uncertainty we are dealing with. It also breaks down a complicated input like weather into its constituents, whose effects can be measured on an engineering system. Furthermore, having a mathematically measureable number allows us to use sophisticated tools from probability theory.

Generally, uncertainty about a quantity is expressed in terms of statistics, for example, mean value and standard deviation. But to get a complete description of uncertainty, one needs the probability distribution function (pdf) of the random variable. Using only the mean and standard deviation provides all the relevant information about a distribution only if the random variable is normally distributed. Normal distributions therefore enjoy favor among engineering practitioners. Other distributions, which sometimes are harder to define and work with numerically, have advantages in terms of flexibility in properly fitting different types of data, for example, the beta distribution. The choice of correct distribution is critical for modeling uncertainty. Indeed, many engineering disasters could be averted if the normality assumption was not used indiscriminately.

For the sake of clarity and mathematical rigor, unless otherwise stated, we will denote the decision as choosing a value of the decision variable $\mathbf{x}$, which could be a vector or a scalar. The objective of a decision is to maximize a utility function or its expectation when uncertainty is present. This utility function is not always known, but the premise in decision analysis is that such a function exists for every decision maker for a given situation and can be elicited using properly framed questions. If the uncertainty is present in the variable $\mathbf{x}$ itself, we cannot control its realizations. In such cases we control its statistics, for example, the mean or variance. The contribution to the uncertainty can also come from sources beyond our control and not necessarily embedded in $\mathbf{x}$. Here are some examples of uncertainty affecting engineering decision making:

1. In metal cutting there is usually a tolerance associated with the target length to be cut. Example: In a machine shop, cutting target shaft lengths of 10 cm results in an error of ±2 mm. The design variable here will be the mean or nominal length.
2. Material properties are not always fixed. Example: A composite material can have a normally distributed tensile strength with mean equal to 60 GPa and a standard deviation of 2 GPa.
3. The price of an end product is tough to determine in the beginning of the design cycle. Complex systems such as automobiles can be priced using various methods leading to significant uncertainty. Example: A concept car, when finally put into mass production, is going to be priced between $35,000 and $55,000.
4. Weather affects the outcomes in agriculture. Usually these uncertainties are represented as probability distributions over temperature or rainfall or some other measurable quantity. Example: Temperatures below 20°F will destroy a grape crop; we may be interested in finding the probability of this happening.
5. User inputs are considered random variables. Loading of a truck or the terrain it is going to be driven on is usually the end user's choice. The uncertainties in these therefore need to be accounted for in the design of the suspension.

6. Earthquake-induced loads and wind loads in structures are considered stochastic. There is usually some data available in the form of extreme value distributions that are used to determine the probability of whether a structure will be able to withstand the loads over a specified period of time.

3.2.1 Five Rules of Actional Thought

Formal studies in decision making (or for that matter, any mathematical field) have relied on first creating a set of axioms that the decision maker agrees with and expects his or her decisions to follow. These axioms are statements that, while they cannot be proven rigorously, are hard to dispute. An example of an axiom is that if someone is present in Ohio at a particular time, he or she cannot also be in Ankara, Turkey, at the same time. The premise behind an axiomatic decision-making method is that a decision inconsistent with that resulting from the use of the axiomatic approach can be traced back to violation of one of the axioms that are at the heart of that method.

The axiomatic approach to making decisions was first presented by von Neumann and Morgenstern in 1944, in their classic book *Theory of Games and Economic Behavior*. The axioms implied a maximum expected utility criterion for making decisions. They showed that if someone agrees with those axioms, a utility function exists and the alternative that maximizes its expectation results in the best decision. The utility function may not always be used by the decision makers in making decisions; we may not even be able to elicit the function—their result simply shows that such a function exists nevertheless. These axioms have seen some slightly different implementations over the years, for example, by Savage (1954), Keeney and Raiffa (1994), Nikolaidis et al. (2011), and Howard (1988, 2007). In this book we will follow the five rules of Ronald Howard for their simplicity and ease of application. Other methods exist too, such as fuzzy logic (Zadeh, 1996) and analytical hierarchy process (Saaty, 1990), but we refer the reader to other sources for these methods.

The five rules set the ground rules that will help us develop a methodology for making decisions under uncertainty. They are order rule, probability rule, equivalence rule, substitution rule, and choice rule. These rules are hard to dispute and are enough to make normative decisions under uncertainty involving not only one, but multiple attributes. These rules are also general enough that value of information studies can be performed as well. Any inconsistencies can be traced back to one of the rules (axioms) being violated and the decision maker should be aware of that. Furthermore, the rules address only actional thoughts, that is, thoughts that will result in an action.

3.2.1.1 *Order Rule*

The order rule states that a decision maker, when faced with deterministic outcomes, can rank these outcomes in the order of his or her preference, and

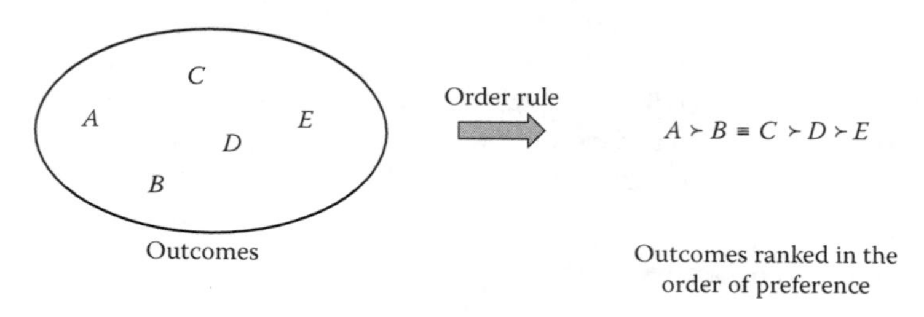

FIGURE 3.1
The order rule ranks outcomes in the order of preference of the decision maker.

ties are allowed (Figure 3.1). This rule ensures transitivity in the preferences of outcomes. For example, if a decision maker has five outcomes to consider, A, B, C, D, and E, one of the ways he or she can rank them is $A \succ B \equiv C \succ D \succ E$. The sign $\succ$ signifies "is preferred to" and the sign $\equiv$ signifies "is equally valuable as." So, for the example mentioned, the decision maker prefers option A to B, is indifferent between B and C, prefers C to D and D to E. This automatically implies that A is preferred to C, D, and E (transitivity). Transitivity will also imply that B and C are preferred to E.

3.2.1.1.1 Overcoming Resistance to Order Rule

Decision makers may resist providing a preference order for two reasons: because they are *unsure of the worths* of the outcomes or because they think there is *intransitivity* among the outcomes. The role of the facilitator is to make the decision maker aware of the repercussions of not providing a preference order. If decision makers are unsure of the worths of the different outcomes to them, then they should consider them equally desirable until further investigations are made. More often than not, collecting more information about relative worths of outcomes is justified, because implementation of the order rule is one of the first steps in decision analysis, and the rest of the steps depend on a properly defined preference order. In simple words, this step identifies what is wanted and is the basis of any decision making.

Lack of transitivity exposes decision makers to indecision, or worse yet, they become vulnerable to someone taking advantage of them. The indecision can be exploited using what is called the *money pump* argument. Imagine being indecisive about purchasing a mobile phone from a store. More specifically, let us say that options A and E represent mobile phones, and D is the mobile phone you currently own. Let us now introduce intransitivity in your preference order, as shown in Figure 3.2. The intransitivity is introduced by including the step where E is preferred to A. As is evident, this leads to a cycle. The store can have you buy option A because you prefer it to your current phone. Then it can make you exchange that for option E, and finally for your own phone. The store (if its preference order satisfies transitivity) can also take advantage of this situation by suggesting something that increases its profit. Or it can charge you fees every time you exchange phones. It is

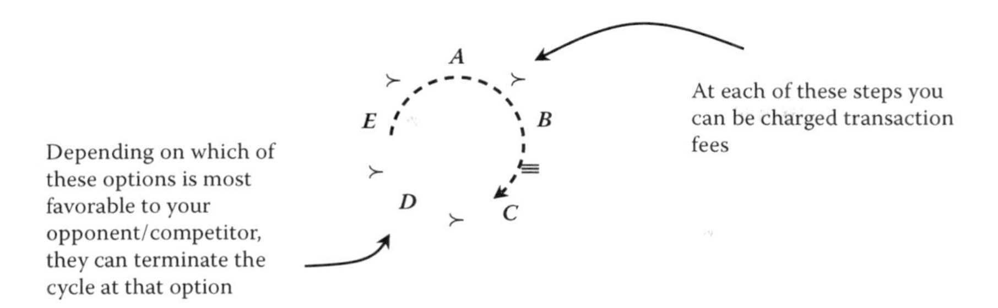

FIGURE 3.2
Intransitive preferences leave you vulnerable to being taken advantage of—the money pump argument.

evident that having intransitivity in preferences is to be avoided. A facilitator who is assessing the decision maker's utility function can use the money pump argument to convince the decision maker to think deeply about his or her preferences and avoid intransitivity. The reader might argue that intransitivity is obviously wrong, and therefore he or she will never exhibit it in his or her preferences. In complex engineering decision problems it is seen that intransitivity is often inadvertently introduced, particularly when dealing with outcomes involving multiple attributes. When a mathematical model of the DM's preferences is being handled by the computer, this intransitivity can lead to grave errors in decision making.

In group decision making it is often observed that pairwise comparison of alternatives can result in loss of transitivity. There are other impediments to group decision making, as necessitated by Arrow's impossibility theorem, as we discussed in Chapter 2. In this chapter, though, the focus is on one decision maker only.

3.2.1.2 Probability Rule

The probability rule states that a decision maker can assign a probability to an outcome based on his or her degree of belief about the likelihood of its realization. At first, this rule appears controversial. People generally argue that they cannot assign probabilities to events about which they have limited knowledge. This can be alleviated by understanding that probability is a degree of belief. Probability is a subjective construct and is the degree of belief about the likelihood of an event in the decision maker's mind. For example, let us say that one is asked: What is the probability that the weight of a certain automobile is greater than 800 kg? If we are totally oblivious to vehicle weights, we can assign a probability of 0.5 (an unbiased value). If we are allowed to look at the automobile, we can compare it with the weight of something we know, for example, the vehicle we own. Finally, depending on what is at stake, we can do investigations to find the true weight of the automobile. It is clear that at each of these steps we had *some* idea of the weight of the automobile. *A decision made using our current state of knowledge can be a good decision if collecting further information is cost-prohibitive or is precluded because of other reasons, such as time pressures.*

Change in degree of belief is brought about by analysis or by collecting more information. In Chapter 6, we will cover different methods of encoding our degree of belief about the likelihood of an event. If one were to make a decision in the presence of a substantial amount of uncertainty, one could first make a meta-decision as to whether it is worth first collecting more information. Depending on the source and quality of information, such a determination can be made. We will cover this in detail when we discuss value of information later in this chapter. The bottom line here is that we can always assign a probability to an outcome, based on our current state of knowledge.

3.2.1.2.1 Overcoming Resistance to Probability Rule

If the probability rule is violated, we cannot encode the decision makers' degree of belief regarding the likelihood of events. The general misconception about this rule that one needs a lot of information to assign probabilities is generally not true, as we discussed above. At every stage of information one has the ability of assigning a probability to an outcome and making a good decision. Such situations arise all the time in real life. The role of the facilitator is to make the decision maker comfortable with the notion of probability as a degree of belief based on the current state of information. A meta-decision to acquire more information can always be made.

3.2.1.3 Equivalence Rule

The equivalence rule states that a decision maker is indifferent between a sure outcome and an uncertain lottery involving best and worst outcomes, for a certain probability distribution over the outcomes of the lottery. This can be easily explained using the earlier example. Let us say that you are given a choice between the sure outcome B and a lottery between options A (best outcome) and E (worst outcome). Let us also assume for clarity that A is worth \$100 to you, B is worth \$10 to you, while E is worth \$0, as in Figure 3.3. Option 1 consists of choosing the sure amount B, and option 2 consists of choosing the lottery, where you have the possibility of getting the best option A with a probability p or getting the worst option E, with probability $1 - p$.

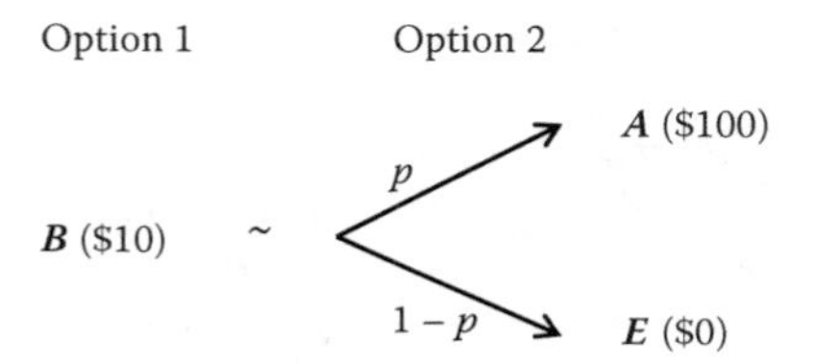

FIGURE 3.3
Establishing equivalence between a sure outcome and an uncertain lottery.

The value of p is a variable and is called your *preference probability* associated with outcome B because it is a measure of how much you value B with respect to A and E. Clearly the higher the value of p, the more you value B. The idea is that it would require the lottery to be really lucrative, that is, give you a very good chance of getting the best option, A, for you to give up the sure option, B. Similarly, a low value of p shows that you value B less. There is also the issue of risk averseness, which we will get to in the later sections, but roughly speaking, the higher the value of p, the higher your aversion to risk as well. It should already be becoming clear that your decision-making behavior under uncertainty is already being investigated using the decision analytic approach.

Let us get back to the decision. It is clear that for $p = 0$ you would prefer the sure amount (option 1) because the decision is a choice between a sure \$10 and an effectively sure \$0. Similarly when $p = 1$, you would prefer the lottery (option 2) because the choice is now between a sure \$10 and an effectively sure \$100. It follows directly that there must exist a probability between 0 and 1 where you will be indifferent between the two options. The beauty of this step is that the *decision maker chooses* the preference probability that suits his or her own preferences, thereby providing insights into his or her decision-making behavior under uncertainty.

3.2.1.3.1 Overcoming Resistance to the Equivalence Rule

If the equivalence rule is not satisfied, this means that the decision maker cannot provide a probability where he or she will be indifferent between the sure amount and the lottery. Indeed, he or she agrees that at the extreme values of the probability he or she can make a choice with certainty. Showing no inclination to provide a preference probability indicates that the decision maker will be highly inconsistent in his or her decisions over time. As an example, he or she might choose option 1 when the probability p is 0.9 and option 2 when $p = 0.6$, which is clearly inconsistent. This also exposes him or her to being taken advantage of, as we shall see below, where the decision maker even shows a lower degree of imprecision in his or her preference probability than the example above.

Let us assume that David is only somewhat indecisive and says that he is indifferent between the lotteries for the value of p between 0.4 and 0.6. In other words, for $p < 0.4$ he would definitely prefer the sure amount, and for $p > 0.6$ he would definitely prefer the lottery. For $0.4 \le p \le 0.6$ he is undecided and sometimes prefers the lottery, and sometimes prefers the sure amount. It is clear, therefore, that he can at least sometimes exchange a lottery with associated probability p with a lottery with associated probability q, such that $p < q$, so long as $p, q \in [0.4, 0.6]$. This is clearly going to result in sure loss because the desirability of a two-outcome lottery like the one shown in Figure 3.3 increases with increasing associated probability. He will miss moneymaking opportunities in some cases, while in others he will rue his decision to take the risk. The inconsistencies in application of the equivalence rule sometimes open up arbitrage opportunities in stock options trading.

In real-life situations where decisions under uncertainty can at best be represented with compound lotteries with many outcomes, it is easy to be unaware of these subtle inconsistencies in one's preferences. This argument will be clearer once we talk about utility functions where we show that the utility of an outcome is simply the preference probability of a lottery between the best and worst outcomes. As a result, if one has a strict preference order over an attribute (e.g., money), one would provide a single value of the probability as opposed to a range.

Making decision makers cognizant of these situations will go a long way in convincing them that they must provide a single number for the preference probability. It will not only help them to remain consistent over time between decisions, but also protect them from others taking advantage of them. Furthermore, a decision maker trained in providing preference probabilities understands the criticality of the step in encoding his or her preferences.

3.2.1.4 Substitution Rule

The substitution rule states that when equivalence has been established between a sure outcome and a lottery between the best and worst outcomes, the decision maker should be indifferent when one is replaced with the other. In other words, for the preference probability given by the decision maker, as described in the previous step, he or she is always willing to substitute the lottery for the sure amount, and vice versa. This rule ensures that the decision maker truly believes in the preference probabilities and will not deviate from his or her preferences in practice. Moreover, having this rule guarantees that complex lotteries between outcomes can be reduced to a simple lottery between the best and worst outcomes.

Figure 3.4 shows how a lottery can be converted into a compound lottery containing only the best and worst options. After simplification, the compound lottery becomes just a binary lottery. Different compound lotteries, corresponding to different decision alternatives, can then be compared. Realize that each alternative in a decision situation will result in a different binary lottery.

3.2.1.4.1 Overcoming Resistance to the Substitution Rule

If the decision maker thinks that the substitution rule does not hold, then the preference probability values given in the previous step are not true. A decision maker must truly provide the preference probability values and act upon them. Not allowing substitution means that a decision cannot be made. It is possible that the facilitator may have to revisit the equivalence rule step with the decision maker to get more conservative preference probabilities.

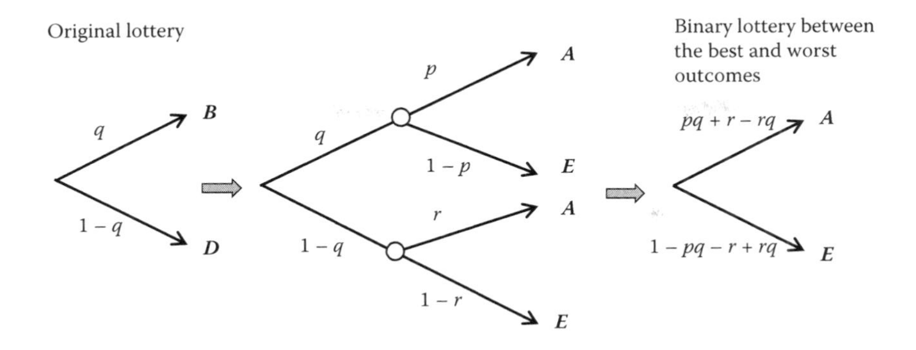

FIGURE 3.4
The decision maker provides a preference probability p for option B and r for option D, and acts on it. The two lotteries (original and the binary lottery on the right) become equivalent.

There are other reasons for the decision maker's indecision as well. It is possible that the worths of different outcomes have not been clearly identified by the decision maker. This would also be a reason for the facilitator to go back and redo the steps for the first three rules.

3.2.1.5 Choice Rule

Let us say that you are faced with many alternatives, each involves a lottery between outcomes (e.g., the original lottery in Figure 3.4). Notice that the original lotteries do not have to be binary; they may involve many outcomes. Regardless of how complicated they are, using the previous rules you would have converted all of them into a binary lottery between the best and worst outcomes. The choice rule states that once all alternatives have been converted into binary lotteries, each between the best and worst outcomes, the alternative that is chosen is the one that has the highest probability associated with the best outcome. In other words, this probability is a measure of the desirability of the alternative. So the decision maker effectively chooses the alternative with the highest probability of giving the best outcome. This rule is pretty obvious because all the alternatives are now represented using binary lotteries with the same outcomes, just different probabilities.

3.2.1.5.1 Overcoming Resistance to Choice Rule

It can be easily shown that if the decision maker agrees with the first four rules and the probability calculations, and has a utility function that is increasing in the desirability of the outcomes, the expected utility will be highest when choosing the alternative with the highest probability associated with the best outcome. A decision maker is making an irrational decision if he or she chooses any other option.

Example 3.1

Kendra ranks five alternatives as follows. Is she consistent with the five rules?

a. $A \succ B \equiv C \succ D \succ E$
b. $A \equiv B \equiv C \equiv D \equiv E$
c. $A \succ B \succ A \succ E \succ C \succ D \succ E$
d. $A \succ B$, $B \succ D$, $D \succ C$, and $C \succ E$
e. $A \succ B$, $C \succ D$, and $C \succ E$

SOLUTION

a. Yes, because she provides a complete transitive ordering.
b. Yes, she simply prefers them equally.
c. No, her preferences are not transitive. She prefers A to B and then B to A. Also she prefers E to C to D, but ends up preferring D to E.
d. Yes, the letters are in a different order and she provides pairwise preferences, but they are consistent.
e. No, because she does not provide complete ordering of the outcomes. For example, it is not known if she prefers A to C. Similarly, it is not known if she prefers D to E, and so on.

Example 3.2

For case (a) in Example 3.1, Kendra provides the following preference probabilities for the intermediate options when asked to compare them with a lottery between the best option (A) and the worst option (E). Are they consistent?

a. $P_{pref}(B) = 0.7$, $P_{pref}(C) = 0.6$ and $P_{pref}(D) = 0.2$, and
b. $P_{pref}(B) = 0.6$, $P_{pref}(C) = 0.6$ and $P_{pref}(D) = 0.7$, and
c. $P_{pref}(B) = 0.3$, $P_{pref}(C) = 0.3$ and $P_{pref}(D) = 0.2$, and
d. $P_{pref}(B) = 0.99$, $P_{pref}(C) = 0.99$ and $P_{pref}(D) = 0.985$, and
e. $P_{pref}(B) = 1$, $P_{pref}(C) = 1$ and $P_{pref}(D) = 0$, and

SOLUTION

a. No, because her preference order tells us that she prefers B and C equally. Her preference probability assignments, on the other hand, favor B.
b. No, because her preference order tells us that D is less desirable than B and C.
c. Yes, because she assigns equal preference probabilities to B and C, and this value is greater than that for D.
d. Yes, because she assigns equal preference probabilities to B and C, and this value is greater than that for D. She just appears to be extremely risk-averse.

e. No, because a preference probability of 1 means that the options are as desirable as the best option, which is A in this case. Similarly, a preference probability of 0 means that an option is as desirable as the least desirable option, which is E in this case. Both of these disagree with her preference order.

3.2.2 Role of a Facilitator

The role of a facilitator is very critical in decision analysis, especially engineering design decision making. Designers and people in general are averse to being told how to make decisions. They usually confuse the decision analytic methods with something that encroaches upon their personal preferences. One of the central roles of the facilitator is to make the decision maker realize that decision analysis does not question the preferences of the decision maker (and thus can be considered amoral). What decision analysis does do is help the decision maker make decisions consistent with his or her preferences. The preferences, once properly encoded in a utility function, can be used to normatively make decisions in situations in which the decision maker will not be to able make decisions due to the effort required. The maximization of the expectation of utility can be automated using any of the optimization techniques suiting the properties of the problem.

The facilitator should always be able to convey to the decision maker that every mathematical manipulation, whether it is made by a person or in an automated fashion on the computer, is justified. Furthermore, any discrepancy between the decision arrived at in an automated environment (e.g., through optimization) and what the decision maker wants to do can be reconciled by pointing out to the decision maker that he or she is violating one or more of the rules that he or she agreed to as being rational. If the decision maker is still not convinced, it might make sense to go over the lottery questions again and see whether the decision maker wants to change any of the responses.

3.3 Expected Utility Criterion to Make Decisions

In this section we show how to implement the five rules using the notion of expected utility. A utility function is a representation of the preferences of the decision maker in mathematical form. It assigns a numerical value to the outcomes, thus measuring their desirability. If a utility function is elicited properly, normative decision analysis says that a rational decision maker tries to maximize the expectation of this function when making decisions under uncertainty. Without any loss of generality, we can assume that the utility function of a decision maker is scaled between 0 and 1. A utility value of 0 corresponds to the least acceptable outcome, and 1 corresponds to the best possible outcome.

We will start by showing that maximizing the expected utility criterion is consistent (or can be made consistent) with the five rules we presented in the previous section.

1. Order rule: Utility functions are consistent with the order rule because we can assign different outcome utility values (real numbers between 0 and 1) according to their desirability. For example, if the attribute is money and you prefer more money to less, the utility function is defined as increasing in the amount of money. Transitivity is guaranteed on the real line.
2. Probability rule: Since probabilities of events (outcomes) are independent of their desirability, a decision maker can assign probability to events by themselves or with the help of the facilitator. This rule holds as well.
3. Equivalence rule: This rule is actually used to elicit utility functions so that they are consistent with all five rules. Consider the decision between a certain intermediate option and a lottery between the best and worst outcomes (Figure 3.5).

   The underlying decision made here is comparing and equating the expected utilities of the two options.

$$U(B) = pU(A) + (1-p)U(E) \tag{3.1}$$

   But $U(A) = 1$ and $U(E) = 0$ since they are the best and worst outcomes, respectively; we then have

$$U(B) = p \tag{3.2}$$

   Therefore maximizing expected utility is consistent with following the five rules of actional thought if we assign to each outcome a utility that is equal to the preference probability associated with it.

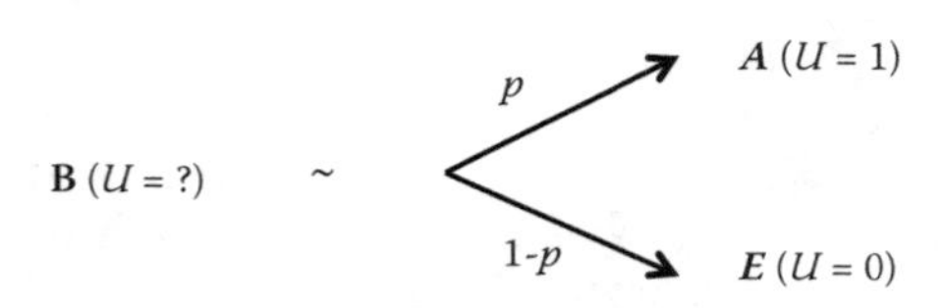

FIGURE 3.5
Using the equivalence rule to assess the utility function.

4. Substitution rule: This rule implies that two lotteries can be substituted for each other if they have the same expected utility. Since the equivalence rule holds, the expected utility criterion will remain consistent with the five rules.
5. Choice rule: Maximum expected utility is achieved only when the chosen alternative has the highest probability associated with the best outcome. Therefore, this rule holds as well.

It follows that we can use the maximum utility criterion to made decisions, and we will be consistent with the five rules of actional thought. The key is to assess the utility function correctly by framing the decision properly and asking well-designed lottery questions.

3.3.1 Problem Framing

Proper framing of decision problems is critical in decision analysis. In general, one does not consider all the possible outcomes when talking about a decision. For example, a microwave oven manufacturer does not consider the possibility of demand affected by a catastrophic nuclear disaster when choosing the range of colors to offer. Similarly, an automobile manufacturer does not worry about solar winds when introducing a new model. Needless to say, since the two extreme events happen with a nonzero probability, the effort required to model the effect on the decisions is almost never going to be justified.* Conversely, sometimes decision makers have too narrow of a frame; that is, they do not consider the necessary factors that will have a tangible effect on the decision. Proper framing is therefore critical.

Problem framing is an art as well as a science. Of course a model that determines unequivocally which of all the externalities to consider, as well as their effect on a decision, will be immensely useful. However, implementing it will, by definition, be just as hard as *solving* the decision problem with all the externalities considered. Therefore, sometimes a meta-decision is first made. Using rigorous thought, expert advice, and probability calculations, the decision maker and the facilitator should reach a consensus on the following:

1. What is it that I am trying to achieve? What are the attributes that I want to maximize or minimize?
2. What are the controllable and uncontrollable variables? What level of control do I have on the controllable variables?
3. How does uncertainty affect the outcomes that I am considering?

* Even if these events end up affecting the outcome, it does not necessarily point to a bad decision.

4. Am I considering attributes or uncertainties that will have minimal effect on the quality of my decision?
5. Are there attributes and uncertainties that might have a tangible effect on the quality of my decision that I am not considering?
6. Do I have the resources to make the decision with the frame I am considering? Do I need more resources or people?
7. Do I have the authority to make the decision?
8. Is my decision going to change substantially if I include more externalities?
9. Can I delay the decision? If so, how will this impact my decision?

3.3.2 Utility Functions and Value Functions

Before we get deeper into risk attitudes and the functional forms of utility functions, it is better to pause and understand the difference between value functions and utility functions. Both functions rank alternatives in the order of desirability. The key difference is that a utility function can also be used in the presence of uncertainty; i.e., it helps correctly rank alternatives that may be uncertain. Value functions do not do that; they can only rank deterministic alternatives. Clearly, all utility functions are value functions too, but not all value functions are utility functions. Many times a value function is used to make decisions under uncertainty, which results in grossly incorrect decisions, while at other times significant effort is expended in assessing utility functions where value functions would have sufficed.

An example will help explain this concept. Let us consider the attribute of money. Most people prefer more money to less, and as a result, any increasing function over money will work as a value function, for example, $\log(x)$, $\sqrt{x}$, x^2, e^x, where x is the money amount. But they could not all serve as utility functions. Let us say that you are offered a lottery that pays \$20 with 50% probability and \$5 with 50% probability, and the lottery costs \$12. Will you buy this lottery? Table 3.1 shows the expectations of the four

TABLE 3.1

Inconsistent Decisions Using Value Functions Showing How They Cannot Always Be Used to Make Decisions under Uncertainty

Value Function	Expectation of the Value Function for the Lottery	Value Associated with \$12	Purchase Lottery?
$\log(x)$	2.303	2.485	No
$\sqrt{x}$	3.354	3.464	No
x^2	212.5	144	Yes
e^x	2.43E+08	162754.8	Yes

value functions we showed, in an attempt to see if we can make decisions using them.

Since all the functions rank deterministic outcomes the same, they could all be used to rank deterministic options. However, as we can see from the table, when uncertainty is present, their expectations cannot be used because they give inconsistent answers. In fact, of all the value functions, *only one* (up to a linear transformation) can be used to make a decision under uncertainty, and it is called the utility function. We will see in the next section how utility functions incorporate uncertainty in a decision. We will also see how some of the functional forms above are parameterized to fine-tune a function to a particular decision maker's preferences.

3.4 Utility Functions and Incorporating Uncertainty in Decisions

Utility functions need to be assessed carefully. There are two ways to do so: by using stated preferences or by using revealed preferences of the decision maker. Using stated preferences requires actively asking relevant questions to the decision maker. We can then populate enough number of points on the utility curve so the function can be interpolated or modeled by fitting a closed-form expression. Using revealed preferences simply means that the decision-making behavior of the decision maker is observed and a utility function fitted to the observed behavior. Acquiring revealed preferences does not involve actively asking questions. While the fitting process is similar, and most of the treatment in this chapter is consistent with both, the two methods have advantages and disadvantages, as shown in Table 3.2. It is generally advised to consider both while assessing utility functions.

TABLE 3.2

Advantages and Disadvantages of Using Stated and Revealed Preferences to Elicit a Utility Function

Advantages	Disadvantages
Stated preferences	
• Relatively easier to acquire • Can be used to determine utility of outcomes that may or may not exist	• No guarantee that the DM will be consistent with his or her stated preferences when making decisions in real life • DM may not even have a preference formed in his or her mind until he or she faces a decision situation
Revealed preferences	
• Reflect the observed true decision-making behavior	• Must be acquired by collecting and analyzing data from previous decisions • Cannot be used if decision scenarios change significantly, such as in evaluating a radically new design for which no prior preference data exist

Before we jump into how utility functions are assessed, we familiarize the reader with risk attitudes and some commonly used utility functions.

3.4.1 Risk Attitude

One of the strengths of formal decision analysis is that it allows for making decisions under uncertainty *using the preferences of the decision maker.* Two decision makers who rank alternatives the same way may not have the same attitude toward risk. The shape of the utility function incorporates these preferences and can be used to determine the risk attitude. Three types of attitudes toward risk are identified:

1. **Risk-averse:** This is by far the most common type of risk attitude encountered. Risk-averse decision makers will value an uncertain lottery less than the expected value of the lottery. The shape of the utility function of a risk-averse decision maker is concave. It directly follows from Jensen's inequality that the expectation of the function will be less than the function at the expected value of the argument. Figure 3.6 shows some examples of utility functions of a risk-averse decision maker.

   The second derivative test can be used to determine if a utility function implies risk averseness. We essentially verify concavity using the second derivative. Therefore, for a decision maker to be risk-averse:

$$\frac{d^2U(x)}{dx^2} < 0 \tag{3.3}$$

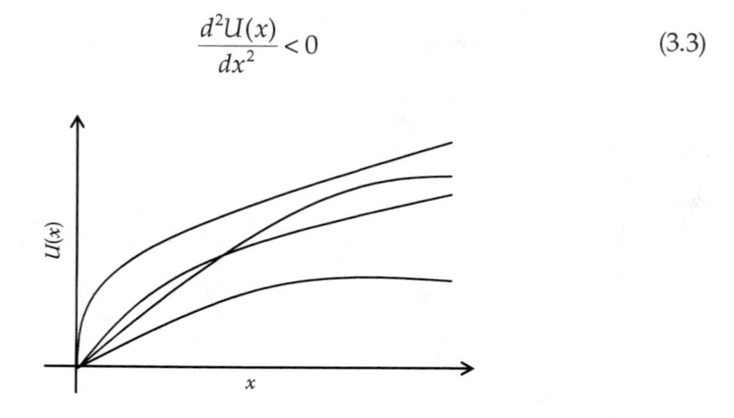

FIGURE 3.6
Different risk-averse utility functions. Note the varying degrees of concavity as well as how it varies along the curve. A decision maker can have different risk attitudes at different points on the utility function.

Extending this to multiple attribute case, risk averseness is determined by whether the Hessian of the utility function is negative definite. The Hessian is given by

$$\begin{bmatrix} \frac{\partial^2 U(\mathbf{x})}{\partial x_1^2} & \frac{\partial^2 U(\mathbf{x})}{\partial x_1\,\partial x_2} & \cdots & \frac{\partial^2 U(\mathbf{x})}{\partial x_1\,\partial x_n} \\ \frac{\partial^2 U(\mathbf{x})}{\partial x_2\,\partial x_1} & \frac{\partial^2 U(\mathbf{x})}{\partial x_2^2} & \cdots & \frac{\partial^2 U(\mathbf{x})}{\partial x_2\,\partial x_n} \\ \vdots & \vdots & \ddots & \vdots \\ \frac{\partial^2 U(\mathbf{x})}{\partial x_n\,\partial x_1} & \frac{\partial^2 U(\mathbf{x})}{\partial x_n\,\partial x_2} & \cdots & \frac{\partial^2 U(\mathbf{x})}{\partial x_n^2} \end{bmatrix} < 0 \tag{3.4}$$

Notice that $\mathbf{x}$ is now a vector given by $\{x_1, x_2, ..., x_n\}^T$ and the inequality < is in matrix sense, in that it implies that the matrix is negative definite. A matrix $\mathbf{M}$ is defined as negative definite if the product $\mathbf{x}^T\mathbf{M}\mathbf{x}$ is less than zero for all nonzero vectors $\mathbf{x}$.

2. **Risk-neutral:** Decision makers who are risk-neutral will value an uncertain lottery at the expected value of the lottery, no more, no less. The utility function of a risk-neutral decision maker is linear and can be represented as straight lines, as shown in Figure 3.7. Decision makers are rarely risk-neutral except for some very rare lotteries involving small sums. A risk-averse or risk-seeking decision maker can be expected to act in a risk-neutral fashion if the range of outcomes is very small.

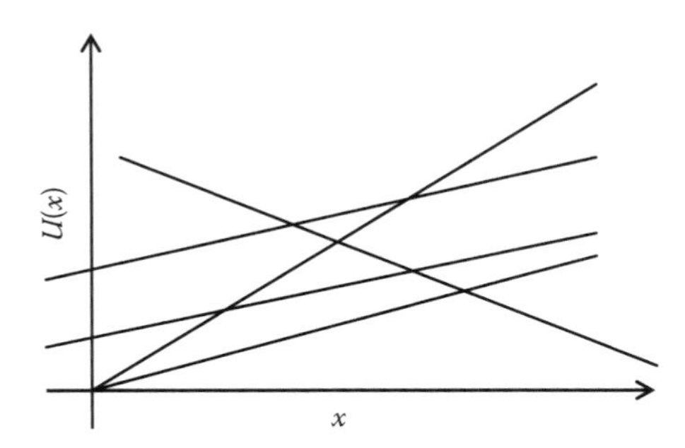

FIGURE 3.7
Different risk-neutral utility functions. The slope and intercept are all different, but all utility functions with a slope of the same sign will lead to the same decisions under uncertainty.

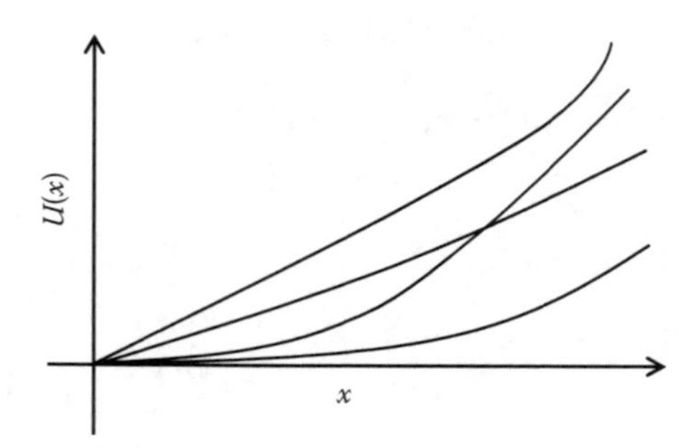

FIGURE 3.8
Different risk-seeking utility functions. Note the varying degrees of convexity as well as how it varies along the curve. A decision maker can have different risk attitudes at different points on the utility function.

3. **Risk-seeking:** Decision makers who are risk-seeking will value an uncertain lottery higher than the expected value of the lottery. Risk-seeking decision makers have a convex utility function; examples are shown in Figure 3.8. As previously, we invoke Jensen's inequality, which in this case implies that the expectation of the utility function is greater than the value at the expectation of the argument.

   The second derivative test can be used to determine if a utility function implies risk-seeking behavior. Just as previously, we essentially verify convexity using the second derivative. Therefore, for a decision maker to be risk seeking:

$$\frac{d^2U(x)}{dx^2} > 0 \tag{3.5}$$

   Extending this to the multiple-attribute case, risk-seeking behavior is determined by whether the Hessian of the utility function is positive definite. A matrix **M** is defined as positive definite if the product $\mathbf{x}^T\mathbf{M}\mathbf{x}$ is greater than zero for all nonzero vectors **x**.

Decision makers can be risk-neutral, risk-seeking, and risk-averse at different points in the utility function. This does not affect the applicability of the five rules of actional thought. It makes definition of risk attitude of the decision maker a little ambiguous in the global sense, though.

3.4.2 Single-Attribute Utility Function

As we discussed earlier, a utility function is a mathematical representation of the preferences of the decision maker. It not only captures the preference order (e.g., more is better) for a particular attribute, but it also takes into account the risk attitude of the decision maker. In this section we will go over some commonly encountered functional forms used for encoding single-attribute utility functions.

3.4.2.1 Linear Utility Function

The linear utility function is simply a linear transformation on the raw attribute. It is of the form

$$U(x) = ax + b \tag{3.6}$$

In single-attribute problems involving no uncertainty, a linear utility function is generally not required because the decision can be made with raw attributes directly. Even when uncertainty is present, the decisions will be based on the expected value of the outcomes and a utility function is not needed. Linear utility functions are, however, useful in scaling attributes for calculations. Linear attribute utility functions are encountered when the outcomes considered fall in a narrow range. An example would be a 50-50 lottery between $10 and $11. Most decision makers will value this lottery at $10.50 or close to it. Another situation where the linear utility function is generally encountered is when the outcomes fall in the lower value range of the attribute under consideration. As an example, we see ourselves not thinking too much before buying lotteries or loaning small amounts to our friends. This is because our utility function is mostly linear for small amounts.

3.4.2.2 Exponential Utility Function

An exponential utility function is the most common utility function encountered in single-attribute problems. The reason is that the exponential function provides for different marginal satisfaction levels at different points on the curve. Since this type of curve is concave, it depicts diminishing incremental satisfaction with increasing level of the attribute. In terms of risk attitude, the exponential function points to a risk-averse decision maker:

$$U(x) = 1 - e^{-x/R} \tag{3.7}$$

For simple decision situations, the exponential function can be assessed using a straightforward technique involving the risk tolerance, R. Let us say that we are assessing the utility function over money for a decision maker. The lottery in Figure 3.9 is presented, and the decision maker is asked to provide the value of x for which he or she is indifferent between the lottery and the sure amount of getting nothing. The lottery consists of him or her getting x dollars with 0.50 probability and losing $x/2$ dollars with probability 0.50.

The expected utility of the sure amount of $0, assuming an exponential utility function, is

$$U(0) = 1 - e^{-0/R} = 0 \tag{3.8}$$

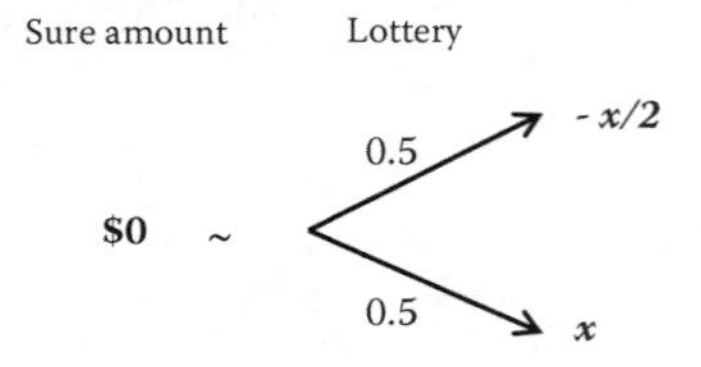

FIGURE 3.9
Lottery to assess the risk tolerance for easy assessment of the exponential utility function.

The expected utility of the lottery, assuming that x indeed equals R, is given by

$$
\begin{aligned}
&0.5U(R)+0.5U(-R/2)\\
&0.5(1-e^{-1})+0.5(1-e^{1/2})\\
&1-0.1839-0.8243 \cong 0
\end{aligned}
$$

Therefore, the two options are equivalent and the lottery does provide a good approximation of risk tolerance R.

In general, we want to be more thorough with the utility function assessment. The above approximation, while technically correct, encodes the utility function based on only one assessment. The assessment error can be minimized if multiple points are assessed throughout the domain of the utility function and a function is fitted.

3.5 Decision Tree

A decision tree is a graphical way to represent decisions, outcomes, and associated uncertainties. It adds to our understanding of a decision by systematically presenting the courses of action. Most decision situations can be represented using decision trees. There are certain conventions used to properly draw a decision tree.

Decision box: A decision node is represented using a box.

Uncertainty oval: An uncertainty is represented using a circle or an oval.

Outcome: An outcome is shown either as a rounded box or in some commercial software as a triangle.

Arrow: An arrow depicts order of execution of a decision tree and does not show dependence.

Figure 3.10 shows the basic elements of a decision tree.

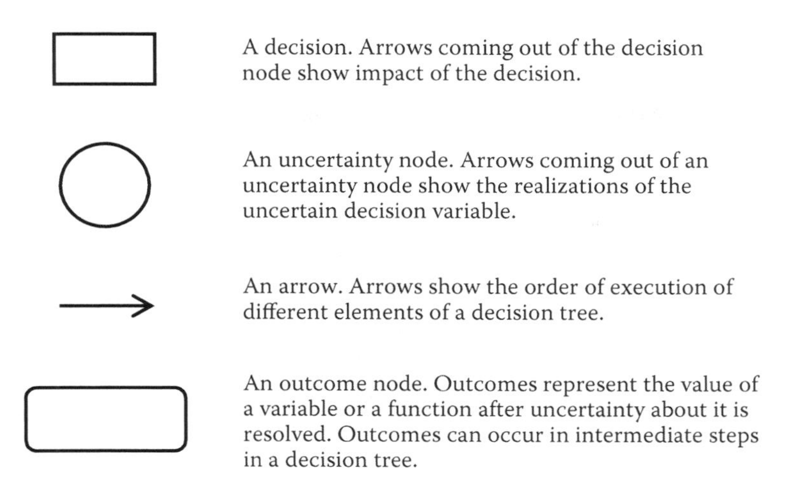

FIGURE 3.10
Graphical elements of a decision tree.

We will show the workings of a decision tree using an example. Consider Conner, a manager in a company department that designs turbine blades. He has two decisions to make: from what material to make the blades, and which of the two design configurations to choose. Table 3.3 shows the material and design choices and the effect of each on the attributes of the problem. Titanium alloy is cheaper but results in a higher probability of failure in operation. Revenue is assumed to be a binary variable with a value of $25,000 when no failure occurs and a value of $15,000 when failure occurs (e.g., $25,000 – warranty costs of $10,000). Conner's utility function is given by $U(x) = 1 - e^{-\frac{x}{15000}}$.

Figure 3.11 shows the decision tree for the turbine blade material and design selection decision. Notice how arrows are used to convey important information about the probabilities and decisions. Since Conner uses the

TABLE 3.3
Conner's Decisions and Their Impact on the Objective

Material Choice	Design Choice	Manufacturing Cost	Revenue
Ti alloy	Design A	$10,000	$25,000 with probability 0.3 and $15,000 with probability 0.7
Ti alloy	Design B	$12,000	$25,000 with probability 0.5 and $15,000 with probability 0.5
Ti-Mb alloy	Design A	$15,000	$25,000 with probability 0.75 and $15,000 with probability 0.25
Ti-Mb alloy	Design B	$16,000	$25,000 with probability 0.9 and $15,000 with probability 0.1

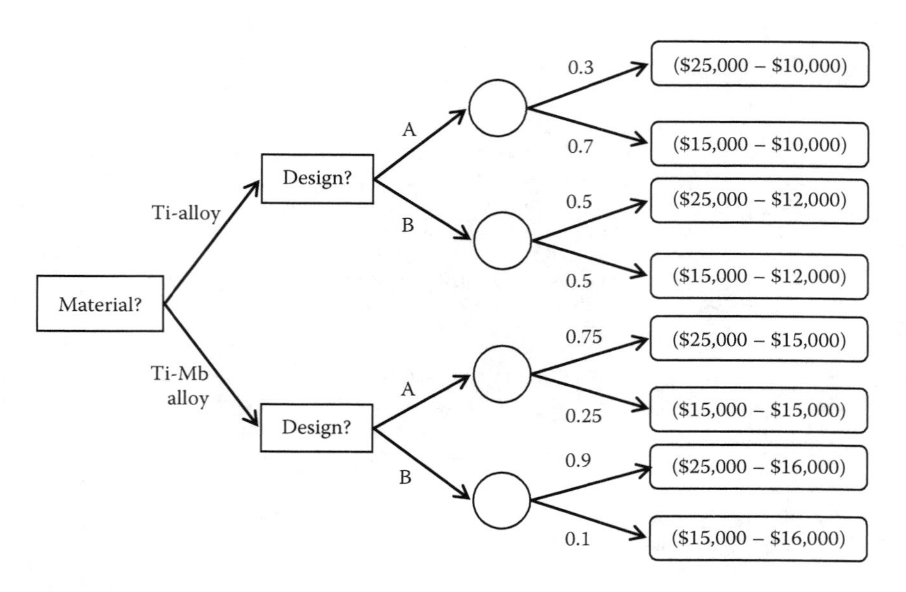

FIGURE 3.11
Conner's decision tree for the turbine blade decision. The outcomes show revenues *minus* the cost of manufacture.

expected utility criteria for making decisions, he has to find the expected utility for the four decision choices. These values are

$$E\left[U(x;Ti,A)\right]=0.3U(15000)+0.7U(5000)=0.388$$

$$E\left[U(x;Ti,B)\right]=0.5U(13000)+0.5U(3000)=0.380$$

$$E\left[U(x;Ti\text{-}Mb,A)\right]=0.75U(10000)+0.25U(0)=0.365$$

$$E\left[U(x;Ti\text{-}Mb,B)\right]=0.9U(9000)+0.1U(-1000)=0.4$$

Therefore, the best design is to choose the titanium-molybdenum alloy and manufacture design B. Notice that this option has a 10% probability of loss, but the high probability of success makes up for it. A lot of real-life decisions are made considering only the worst- or best-case scenario, which is incorrect. Decision analysis helps us make these decisions considering all possible outcomes and in a way that is consistent with our risk-taking behavior. It is not hard to add further decisions and uncertainties in the decision tree. However, the size grows exponentially with the number of "stages" considered because of the tree structure. As a result, decision trees become cumbersome. They are therefore used to convey the main points of the decision situation. Highly complex decision situations can be directly programmed into a computer without resorting to a decision tree.

3.5.1 Sensitivity Analysis

Sensitivity analysis refers to the study of how the optimal decisions change when the parameters of the decision change. Sensitivity analyses can be done on almost any parameter in a decision problem. Consider Conner's decision again. Some of the questions that sensitivity analysis helps answer are: How would the decision change if the revenue values were uncertain? What if the probabilities were different? What if the manufacturing costs were different?

Here we show with an example how sensitivity analysis can be applied to find how sensitive the decision is to changing parameters in a decision problem. Assume that the higher value of revenue from the blades is different from $25,000. In fact, we can vary it within a range, let us say, from $15,000 to $40,000 to see which decisions become optimal. Figure 3.12 shows how the utility from each decision changes as the revenue is varied. We see that design A manufactured with titanium alloy has the highest utility for revenue less than around $24,000. After that, design B manufactured Ti-Mb alloy has the highest utility. The figure visually shows us this transition point. Note that the two decisions (design B, Ti) and (design A, Ti-Mb) never become optimal.

Further sensitivity analysis can also be performed, for example, using the manufacturing costs and the probabilities of each outcome, in a similar way. Visually representing sensitivity information has a lot of value, as it can represent various aspects of the decision problem succinctly. Two-way sensitivity analyses are performed when two parameters are varied simultaneously. In such cases optimal decisions occupy regions in the

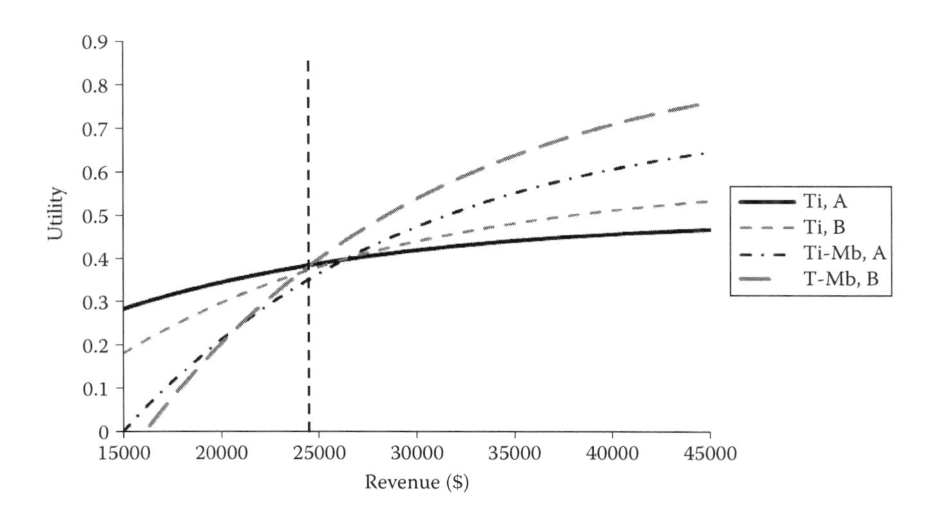

FIGURE 3.12
Sensitivity analysis on revenue in Conner's decision problem.

two-dimensional surface. As the number of variables increase, representing the optimal decision regions becomes harder. Sensitivity analysis is therefore done by first selecting a baseline design. One or two considered are then considered at a time, while setting other parameters constant, to evaluate their effect on the optimal decision.

3.6 Certainty Equivalent of a Decision

Many times, one needs to determine how much a decision situation involving uncertainty is worth in sure amounts. Consider the following examples: A company intends to acquire a license to implement a patented technology. It has to pay the licensing fees up front, even though the outcomes from the use of the technology are uncertain. Oil and mining companies acquire land with uncertainty about the existence and amount of oil or minerals present. Manufacturers routinely invest money in projects where the eventual success of the product and expected revenues are uncertain. In all these cases the decision makers would like to know the sure amount of their decisions under uncertainty.

The sure amount that is equivalent to a decision under uncertainty is called the *certainty equivalent* of the decision, sometimes abbreviated CE. Some authors prefer to call it certainty equivalent *of a lottery*, but we believe that the *lottery* term is very restrictive and can be misleading. Certainty equivalent of a decision is the amount that can make a decision maker give up the decision situation. It is also the amount that he or she is willing to pay to undertake a risky decision. Clearly certainty equivalent is a function of the risk attitude of the decision maker. A risk-neutral decision maker's certainty equivalent for a risky decision is the same as its expected value. This is not true for risk-seeking or risk-averse decision makers.

We now present the way to calculate the certainty equivalent of a decision. The basic premise is that the expected utility from the uncertain decision should be the same as that from the sure amount. That is,

$$E\left[U(x)\right]=U(CE) \tag{3.9}$$

which gives us

$$CE=U^{-1}\left(E\left[U(x)\right]\right) \tag{3.10}$$

Consider the decision tree in Figure 3.13. The decision problem requires a choice between a certain \$50 and a lottery that can pay \$200 with probability

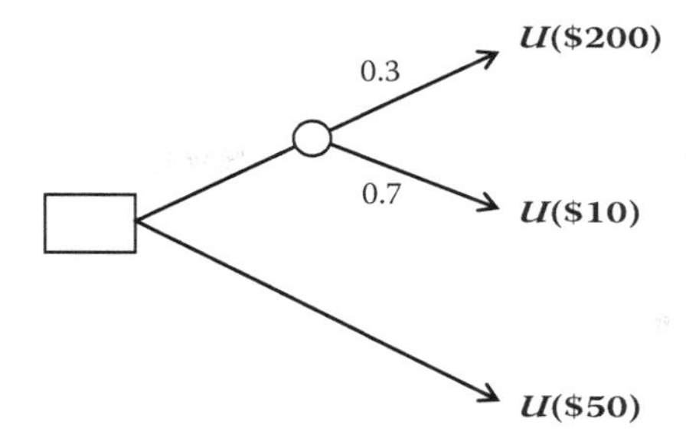

FIGURE 3.13
Finding the certainty equivalent of a decision.

0.3 and \$10 with probability 0.7. Assume that the utility function of the decision maker is given by

$$U(x) = 1 - e^{-\frac{x}{250}} \tag{3.11}$$

Therefore, the utilities of the three outcomes are given by

$$U(200) = 1 - e^{-\frac{200}{250}} = 0.5506$$

$$U(50) = 1 - e^{-\frac{50}{250}} = 0.1812$$

$$U(10) = 1 - e^{-\frac{10}{250}} = 0.0392$$

The expected utility of the uncertain node is given by $0.3 \times 0.5506 + 0.7 \times 0.0392 = 0.1926$, which is greater than that of the deterministic node; therefore, the decision maker would choose the uncertain lottery over the fixed \$50. To find the certainty equivalent of this decision we invert the utility function:

$$CE = U^{-1}(0.1926)$$

or

$$CE = -250 \log(1 - 0.1926) = \$53.5$$

Therefore, the decision is worth \$53.5 to the decision maker. Notice that expectation of the decision is $0.3 \times 200 + 0.7 \times 10 = \$67$, which for a risk-neutral decision maker is also the certainty equivalent. Therefore, the certainty equivalent for a decision is less than the expected value of the decision, as is expected for a risk-averse decision maker.

Now imagine a risk-seeking decision maker faced with the same decision situation whose utility function is given by

$$U(x) = x^2 \tag{3.12}$$

The corresponding utilities are

$$U(200) = 200^2 = 40000$$
$$U(50) = 50^2 = 2500$$
$$U(10) = 10^2 = 100$$

The expected utility of the uncertain node is $0.3 \times 40{,}000 + 0.7 \times 100 = 12{,}070$, which is greater than that for the sure amount, and therefore the decision maker would play the lottery. Notice that we did not need to do any calculations because a risk-averse decision maker also chose the lottery, which implies that a risk-seeking decision maker definitely would. The certainty equivalent of the lottery is given by

$$CE = U^{-1}(12070)$$

or

$$CE = \sqrt{12070} = \$109.86$$

This number is greater than the expected value of the decision ($67). This is to be expected because the decision maker now is risk seeking, and prefers the opportunity to play the lottery over the sure amount with the same expectation.

3.6.1 Delta Property

Howard (1984) defines a delta property in utility functions that is very helpful in calculating the certainty equivalent of a decision when the outcome values themselves are not known with certainty. Consider the above decision situation again. Figure 3.14 shows that scenario where all the outcomes increase or decrease by the same amount. Is there any way to determine the certainty equivalent of the lottery without having to do the calculations again? There is. If the decision maker has a linear or an exponential utility function, the certainty equivalent increases or decreases by the same amount as the outcomes do. This property is known as the delta property.

It is trivial to prove the delta property for a linear utility function. For an exponential function we write

$$U(CE) = 0.3U(200) + 0.7U(10) \tag{3.13}$$

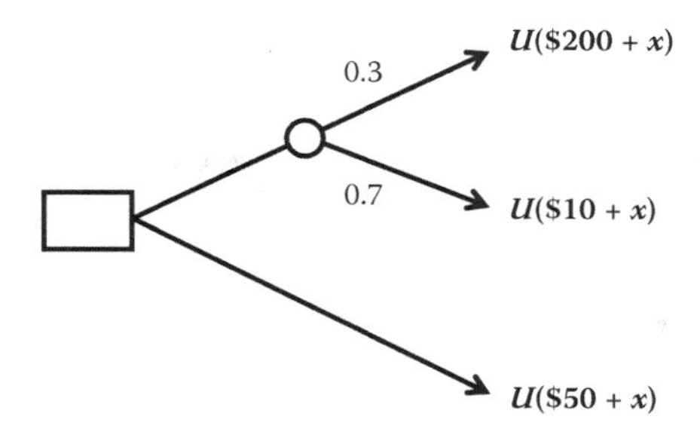

FIGURE 3.14
Finding the certainty equivalent of a decision and delta property.

which we can rewrite for an exponential utility function as

$$1-e^{-\frac{CE}{R}}=0.3\left(1-e^{-\frac{200}{R}}\right)+0.7\left(1-e^{-\frac{10}{R}}\right)$$

Simplifying, we get

$$e^{-\frac{CE}{R}}=0.3e^{-\frac{200}{R}}+0.7e^{-\frac{10}{R}}$$

Multiplying both sides by $e^{-\frac{x}{R}}$, we get

$$e^{-\frac{CE+x}{R}}=0.3e^{-\frac{200+x}{R}}+0.7e^{-\frac{10+x}{R}} \tag{3.14}$$

In other words, the certainty equivalent changes by the same amount as the outcomes.

Example 3.3

Find the certainty equivalent for Conner's turbine blade material and design decision problem.

SOLUTION

In our solution of the decision tree, we showed that Conner chooses Ti-Mb alloy and design B. His expected utility was found to be 0.4. The certainty equivalent is given by

$$CE=U^{-1}(0.4)$$

or

$$CE=-15000\log(1-0.4)=\$7,662$$

So, Conner will give up the turbine decision situation for a sure $7,662. Notice that this is less than the expected value of the decision at $8,000, as expected because he is risk-averse.

3.7 Value of Information

Being able to make good decisions under uncertainty does not mean that we must always contend with it. We could create another decision problem (sometimes referred to as a meta-decision) that first determines if it is worth collecting more information to reduce uncertainty. Recall that we divide uncertainty into two types: aleatory and epistemic. Epistemic uncertainty, which is generally more prevalent than aleatory, can usually be reduced by collecting more information or by seeking expert opinion. Both of these cost resources, usually money and time. Value of information studies seek to determine how much money or resources are worth expending to collect more information. Spending money on information affects the distribution of the outcomes, and unless it does so, it has no value. Information works by modifying the distributions of the random variables such that they get closer to the actual distribution.

Information can be of two types: perfect and imperfect. Understandably, perfect information has higher value than imperfect information. While perfect information can almost never be acquired, its value lies in determining the maximum amount information can be worth. An avenue that promises information at a cost equal to or more than the value of perfect information can be summarily rejected.

3.7.1 Value of Information for Risk-Neutral Decision Makers

We work with a simple decision problem here to explain the concept of value of information. Consider a risk-neutral decision maker confronted with the decision problem of Figure 3.15. The decision is between choosing a sure \$50 or a lottery between \$200 and \$10 with probabilities of 0.3 and 0.7, respectively. Since we have assumed our decision maker to be risk-neutral, he or she would prefer the lottery to the sure amount because its expectation is $0.3 \times 200 + 0.7 \times 10 = \$67$, which is greater than \$50.

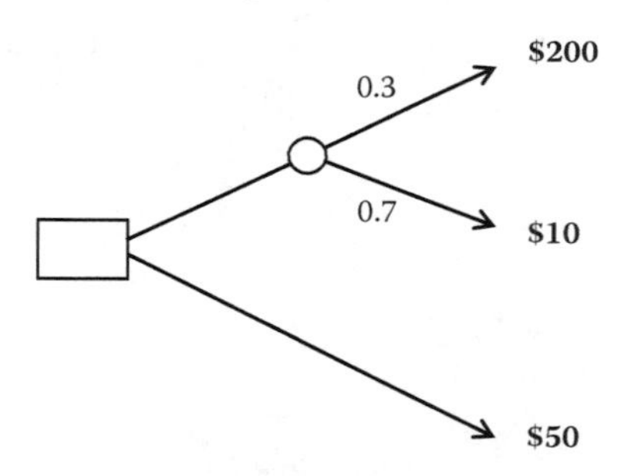

FIGURE 3.15
A decision between a sure amount and a lottery for the value of information example.

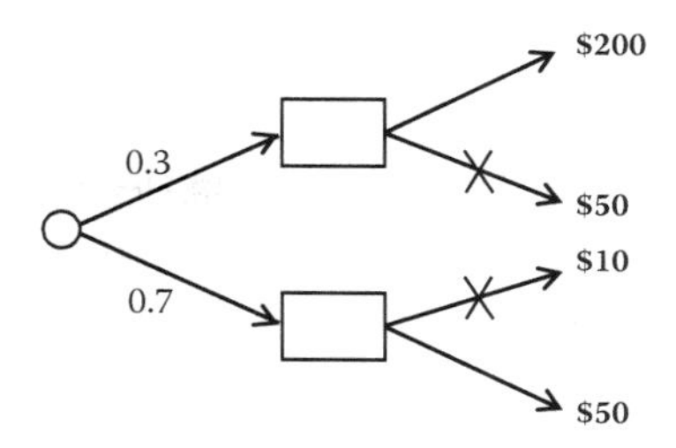

FIGURE 3.16
Inverting the decision tree for the value of information example.

Now imagine that your friend John offers perfect information about the lottery; that is, he can predict the payoff of the uncertain node perfectly. To analyze the situation, we have to redraw the decision problem with the uncertainty node in the beginning. The probabilities of the lottery apply to the information as well;* that is, there is a 0.3 probability that John will say that the payoff is $200, and there is a 0.7 probability that John will say that the payoff is $10. In each case you have a decision between (now) deterministic choices. As shown in Figure 3.16, if the decision maker has a choice between $200 and $50, he or she will choose $200; similarly if he or she has a choice between $50 and $10, he or she will choose $50. As a result, the expected value of the decision is $0.3 \times 200 + 0.7 \times 50 = \$95$. This value is greater than that before the information was acquired. The difference is $95 – $67, or $28. Therefore, the value of perfect information is $28, and this is the maximum amount the (risk-neutral) decision maker should be willing to pay for perfect information to resolve the uncertainty.

We now consider the scenario that that information is not perfect; that is, it modifies the distribution of the outcomes but does not make them deterministic. Consequently, when John now says that the payoff will be $200, it only happens with a probability less than 1 ($10 happens with probability greater than zero). Similarly, when he says that the payoff will be $10, it happens with a probability less than 1 ($200 happens with a probability greater than zero). In this scenario, we need to first calculate probabilities of many intermediate events before we can make a decision. We write the probabilities in shorthand using the following notation:

Probability that the payoff is $200 = P(200) = 0.3
Probability that the payoff is $10 = P(10) = 0.7
Probability that John says that the payoff is $200 = P(J200)
Probability that John says that the payoff is $10 = P(J10)

* Recall that probability is a degree of belief. The probability that John will say that the payoff will be $200 is the same as your initial belief of its being $200, when John provides perfect information.

Probability that the payoff is \$200 when John says \$200 = P(200|J200)
Probability that the payoff is \$10 when John says \$200 = P(10|J200) = 1 – P(200|J200)
Probability that the payoff is \$10 when John says \$10 = P(10|J10)
Probability that the payoff is \$200 when John says \$10 = P(200|J10) = 1 – P(10|J10)

We redraw the decision tree with the above probabilities incorporated in Figure 3.17. Notice that none of the probabilities in the decision tree is explicitly known. Notice also that the decisions are not as straightforward as they were in the case of perfect information; when John says \$200, this does not necessarily mean that the decision maker will choose the lottery over the sure amount. Assume now that you know that when the payoff was \$200, John called it correctly 90% of the time, and when the payoff was \$10, John called it correctly 80% of the time. Therefore, P(J200|200) = 0.9 and P(J10|10) = 0.8.

We first calculate the probability of conditional events. Let us start with P(200|J200). Bayesian probability tells us that it is given by

$$P(200 \mid J200) = \frac{P(J200 \mid 200)P(200)}{P(J200)}$$

Using the total probability theorem we can rewrite the above equation as

$$P(200 \mid J200) = \frac{P(J200 \mid 200)P(200)}{P(J200 \mid 10)P(10) + P(J200 \mid 200)P(200)}$$

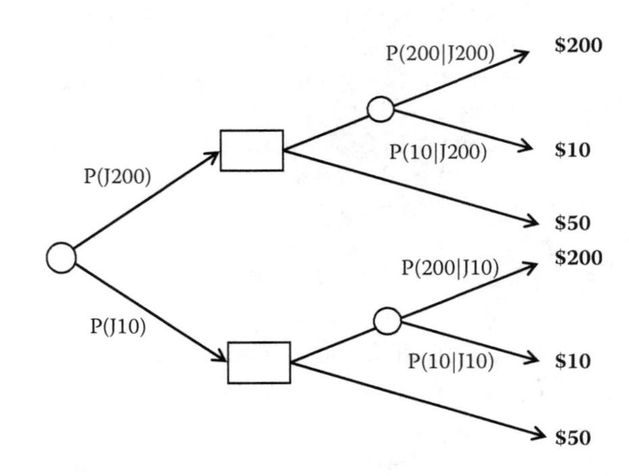

FIGURE 3.17
Updated decision tree after acquiring imperfect information from John.

All the probability values on the right-hand side of the above equation are known; therefore, we have

$$P(200 \mid J200) = \frac{0.9 \times 0.3}{(1-0.8) \times 0.7 + 0.9 \times 0.3} = 0.6585$$

which gives us for the complementary event:

$$P(10 \mid J200) = 1 - 0.6585 = 0.3415$$

Using the Bayesian probability again we have

$$P(10 \mid J10) = \frac{P(J10 \mid 10)P(10)}{P(J10 \mid 10)P(10) + P(J10 \mid 200)P(200)}$$

or

$$P(10 \mid J10) = \frac{0.8 \times 0.7}{0.8 \times 0.7 + (1-0.9) \times 0.3} = 0.9492$$

which gives us for the complementary event:

$$P(200 \mid J10) = 1 - 0.9492 = 0.0508$$

Notice that the probabilities P(J200) and P(J10) have already been calculated when calculating denominators of the two equations. The values are P(J200) = 0.41 and P(J10) = 0.59. We redraw the decision tree with the probability values populated in Figure 3.18.

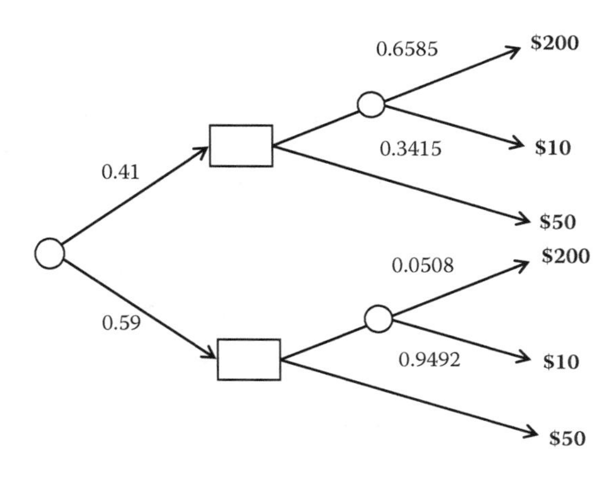

FIGURE 3.18
Updated decision tree after acquiring imperfect information from John with probability values populated.

When John says that the payoff is going to be \$200, the expectation of the lottery is \$135.12, so the lottery should be chosen. When John says that the payoff is going to be \$10, the expectation of the lottery is \$19.66; therefore, the sure amount of \$50 should be chosen. The overall expectation of the decision is 0.41 × 135.12 + 0.59 × 50 = \$84.90. As before, the information increases the expected value of the decision, but by only 84.90 – 67 = \$17.90.

3.7.2 Value of Information Using Utility Functions and Continuous Distributions

We must be aware of the simplifications we made in our examples. We assumed a risk-neutral decision maker, and we assumed that only discrete outcomes are possible. Recall that most decision makers are not risk-neutral. Furthermore, many random variables follow a continuous distribution. As a result, our analysis will change; instead of using the expected value, we will use expected utility *and* we will use calculus to find expected values, to make the determination of how much the information is worth. The general formulation is that value of information (VoI) is found by solving the following equation:

Expected utility without information = Expected utility with information *considering the cost of information*

Or more succinctly, in a general case, solving the following equation gives the value of information:

$$\int U(x) f_X^{without}(x)dx = \int U(x - VoI) f_X^{with} dx \tag{3.15}$$

Notice that information changes the probability distribution of the outcomes, and if it does not, it has no value. Change in distribution can imply just change in the shape of the pdf, or even moving of the support; that is, outcomes that were not possible before become possible now, and conversely, outcomes that were possible before are not possible now. Figure 3.19 shows

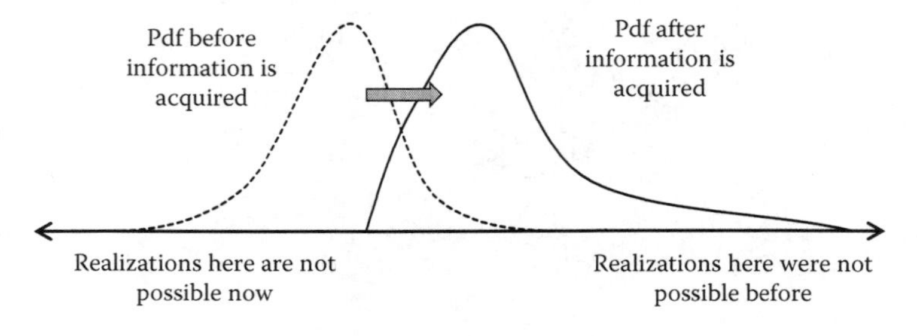

FIGURE 3.19
Information modifying the pdf of the random variable.

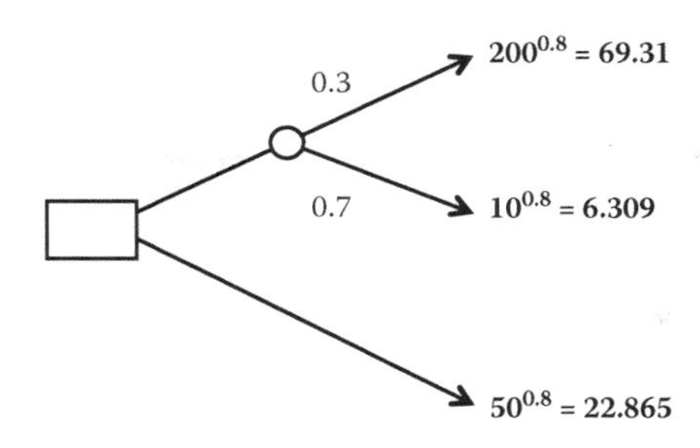

FIGURE 3.20
Value of information using utility functions.

how value of information can modify the pdf of the outcomes. Notice that any shape of the final distribution is possible, not just a shift to the right, as depicted in the figure.

3.7.2.1 Discrete Outcomes

We now revisit our example on value of perfect information, only now considering utility functions. Imagine now that the decision maker's utility function for money is given by $U(x) = x^{0.8}$. Figure 3.20 shows the decision tree.

The expected utility of the uncertain node is equal to 25.21, which is greater than that of \$50 (22.865); therefore, before the information is acquired, the decision maker will choose to play the lottery. When the information is acquired from John, the tree is flipped as before, and the probabilities can be directly used. Value of information (VoI) is given by

$$25.21 = 0.3 \times U(200 - VoI) + 0.7 \times U(50 - VoI)$$

or

$$25.21 = 0.3 \times (200 - VoI)^{0.8} + 0.7 \times (50 - VoI)^{0.8}$$

Numerically solving the above equation, we get $VoI = \$31.86$.

3.7.2.2 Continuous Outcomes

Let us now consider a case where both the uncertainties (the distribution of outcomes) and what the information provider says follow continuous distributions and the decision maker is risk-averse. Figure 3.21 shows the decision tree for the decision problem. The outcome is normally distributed with a

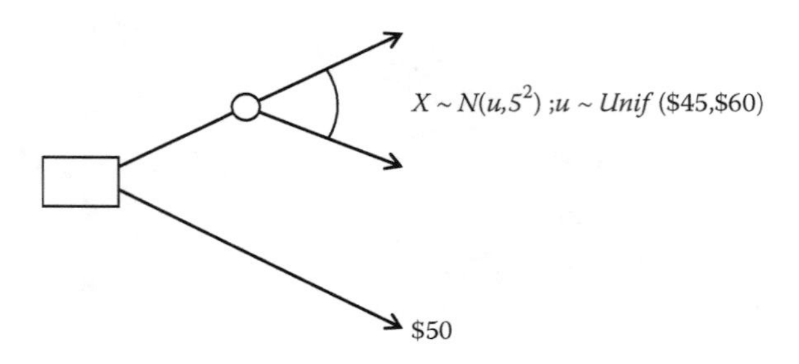

FIGURE 3.21
Value of information using continuous outcomes.

standard deviation of \$5. There is uncertainty about the mean of the outcomes, and it is believed that it is distributed uniformly between \$45 and \$60. The utility function of the decision maker is again given by $U(x) = x^{0.8}$. The expected utility before consulting John is given by

$$E[U(x)] = \int_{-\infty}^{\infty}\int_{45}^{60} x^{0.8} \frac{f_T^N\left(x;t,5^2\right)}{15} dt\, dx \tag{3.16}$$

The expected utility value of the uncertain node is calculated numerically and found to be equal to 23.78, which is greater than the utility for \$50, which is 22.86. Therefore, before the information is acquired, the decision maker prefers the lottery. The decision maker consults John again, who is historically known to provide correct information on the statistics of the distribution. He provides the value of the mean while agreeing that the standard deviation is fixed at \$5. However, he does not provide the actual value of the realization of the outcome, and therefore it still counts as imperfect information. As we have done before, we assume that it is uncertain what mean he will specify. Numerical calculations tell us that when the mean of the normal distribution is less than \$51.5, it is better to go for the certain option. Numerically solving for the value of information gives us a value of \$1.35. Such a low value is expected because John does not resolve the uncertainty significantly.

3.7.3 Value of Information and Reduction of Uncertainty

Information does not necessarily have to reduce uncertainty about an outcome. It should bring us closer to the true distribution of the variable. In other words, if it is more likely than not that the true distribution of the outcomes will be acquired, information has value. Clearly there is no way to know if this will indeed happen. Past experience and track record of the information provider must therefore be judged before committing to compensation. Even in cases where a decision maker is totally ignorant of the true probability distribution, approximations can be made to determine the value of information.

3.7.4 What If the Information Is Incorrect?

Marketing studies and expert opinions are acquired with the faith that they are mostly correct. They usually have a positive value because more often than not they provide good information. In cases where the information they provide is incorrect, they can result in reduced expected utility, further compounded by the fact that the information itself cost money. As previously mentioned, decision makers should carefully look at the track record of the entity providing information before committing to providing compensation. In engineering problems where tests are used to acquire information, the outcomes are pretty much straightforward. If information from tests is found to be wrong, it is sometimes worthwhile to perform another well-implemented test.

There are situations, particularly in large engineering projects such as oil well drilling, where uncertainty affects the outcomes significantly. Furthermore, the information provided by experts (geologists or logging companies in this case) only somewhat resolves the uncertainty. In such cases it is often found that seeking a second opinion may also be helpful. The steps in calculation are the same; every stage where information is acquired changes the probability distribution of the outcomes and improves the expected utility. The amount of money that just compensates for the increase in utility is the value of information.

3.8 Eliciting Single-Attribute Utility Functions

In this chapter we have discussed some commonly used utility functions, how they can be used to make decisions, what the risk attitudes of the decision makers can be, and what the certainty equivalent of a decision means. We also have discussed how to make decisions under uncertainty, given the utility function of the decision maker. So far we have assumed that the utility function is known. We now address the problem of practical assessment (or eliciting) of utility functions. Eliciting utility functions involves asking the decision maker questions which allows the facilitator to encode the decision maker's preferences in a mathematical function.

There are many things a facilitator wants to accomplish while eliciting utility functions. In addition to acquiring the mathematical function, the facilitator should know the decision maker's motivation for undertaking a decision, his or her attributes of concern (for multiattribute problems), and the range of negotiability, and also train the decision maker in providing answers that might help with future assessments. The facilitator must ensure that the decision maker is aware of the decision situation and the decision frame before asking questions. He or she should also check to see if the decision maker is biased or is anchored to providing overly aggressive or pessimistic values.

It is also important to make the decision maker as comfortable as possible during the elicitation process. The utility function elicitation process involves asking many questions that may overwhelm the decision maker. The questions should seem pertinent to the decision problem. To minimize the number of questions asked, each question should be such that it reveals something new about the utility function. Some decision theorists believe that priming the decision maker with lottery questions (which may or may not have to do with the decision at hand) is also helpful. Finally, the facilitator should not be fixated on a particular functional form of the utility function; he or she should select the one that best fits the decision maker's responses.

3.8.1 Range of Negotiability

We have discussed proper problem framing to limit the scope of a decision situation. In terms of attributes, range of negotiability is akin to a proper decision-making frame. Range of negotiability refers to the range of values of an attribute that the decision maker is *actively* considering. For continuous attributes it is a range (or collection of ranges) on the real line, and for discrete variables it is the set of all possible values the attribute can realistically take. For most decision-making situations the range of negotiability is the set of all values between a low and a high value of the attribute. When maximizing an attribute, the lower limit of the range refers to the *minimum acceptable* (x_L) level of the attribute, while the higher value refers to the *maximum reasonably achievable* (x_H) level of the attribute. There are two reasons for not making the range $[x_L, x_H]$ very wide, despite the fact that it would include, in theory, preferences over a wider range of possibilities. First, it makes the utility function very hard to assess, as preferences over a wide range need to be encoded. Finding a functional form that accomplishes that is hard. Second, the range may include values of the attributes that are not realizable in practice, which would result in wasted effort in utility function assessment.

An example can help explain this concept. When you go to buy a car, you do not expect it to give a fuel economy of 150 miles per gallon (mpg) in normal driving conditions. Most people will be reasonably satisfied with 35 mpg. Similarly, most people expect at least 20 mpg from a daily driving car. Conventionally, we would make the utility for 20 mpg equal to zero and the utility for 35 mpg equal to 1. This process is called normalizing (see Figure 3.22). It is clear that your satisfaction will keep increasing beyond 35 mpg and keep decreasing below 20 mpg, but the change is going to be minimal (think slope!). Recall that normalizing the utility function in

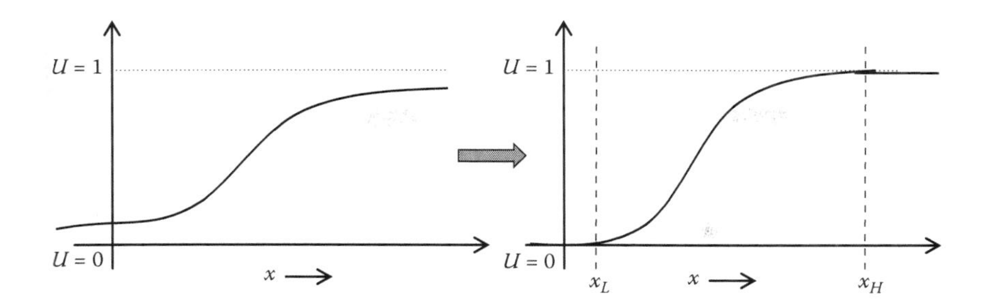

FIGURE 3.22
Normalizing the utility function within the range of negotiability.

the range of negotiability will not change decisions because it is a linear transformation.

As a final note on proper choice of the range of negotiability, when we move to multiple attributes, excessively wide ranges become even more problematic. Independence conditions, which we will talk about later in the chapter, are harder to satisfy. A wide range of negotiability also increases the burden for optimizers when finding the optimal solutions since they have to search a larger search space. Gradient-based optimizers that depend on properties such as convexity and Lipschitz continuity suffer the most because these properties now have to hold for a large range of variable values. In conclusion, the importance of a proper range of negotiability cannot be overemphasized.

3.8.2 Certainty Equivalent Method

We have encountered the CE method in some of our discussions. The certainty equivalent method requires presenting a hypothetical decision situation to the decision maker. On the one hand, there is a lottery, and on the other, a sure amount, which the decision maker provides. The decision maker is asked to provide the value of the sure amount that will make him or her indifferent between the two. The CE method therefore enforces the equivalence rule of actional thought.

Consider the lottery given below (Figure 3.23) for finding out the utility function over a car's fuel economy. If the decision maker chooses the lottery, he or she can get a car that gives 35 mpg with probability 0.4 or one that gives 20 mpg with probability 0.6. Alternatively, he or she can choose a sure value (depicted by x in Figure 3.23), everything else remaining the same. The decision maker has to provide this value of x that will make him or her indifferent between the lottery and x for sure.

The CE method requires a choice of a functional form. The parameters can then be found by solving the following equation:

$$U(x) = 0.4U(35) + 0.6U(20) \tag{3.17}$$

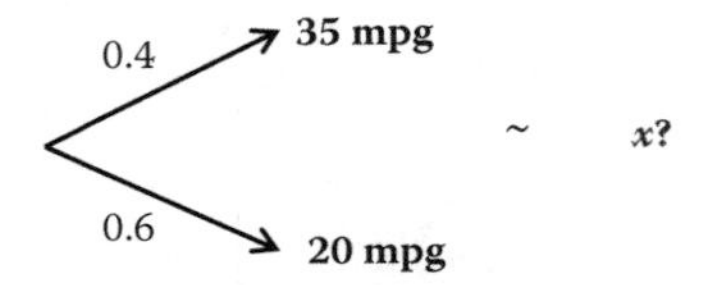

FIGURE 3.23
Utility function assessment using the certainty equivalent method.

For an exponential utility function, we will solve

$$1-e^{-\frac{x}{R}}=0.4\left(1-e^{-\frac{35}{R}}\right)+0.6\left(1-e^{-\frac{20}{R}}\right) \tag{3.18}$$

or

$$e^{-\frac{x}{R}}=0.4e^{-\frac{35}{R}}+0.6e^{-\frac{20}{R}} \tag{3.19}$$

which can be solved numerically. In case the range of negotiability was indeed between 20 and 35 mpg, we can modify the above equation thusly:

$$\frac{e^{-\frac{x-20}{R}}}{\left(1-e^{-\frac{35-20}{R}}\right)}=0.4\frac{e^{-\frac{35-20}{R}}}{\left(1-e^{-\frac{35-20}{R}}\right)}+0.6\frac{e^{-\frac{20-20}{R}}}{\left(1-e^{-\frac{35-20}{R}}\right)}$$

Notice that the value of the upper limit on fuel economy (35 mpg) does not affect the value of R. This is because the expression in the denominator simply scales the utility function within the range of negotiability. The above equation can be simplified as follows:

$$e^{-\frac{x-20}{R}}=0.4e^{-\frac{15}{R}}+0.6$$

For example, the given lottery is $x = 22$, the value of R becomes 4.04338, and the utility function is given by

$$U(M)=\frac{1-e^{-\frac{M-20}{4.04338}}}{1-e^{-\frac{15}{4.04338}}}=1.025-1.025e^{-\frac{M-20}{4.04338}} \tag{3.20}$$

Notice that all the elements of the lottery, such as the probabilities and outcomes, can be changed to assess the utility function at different points

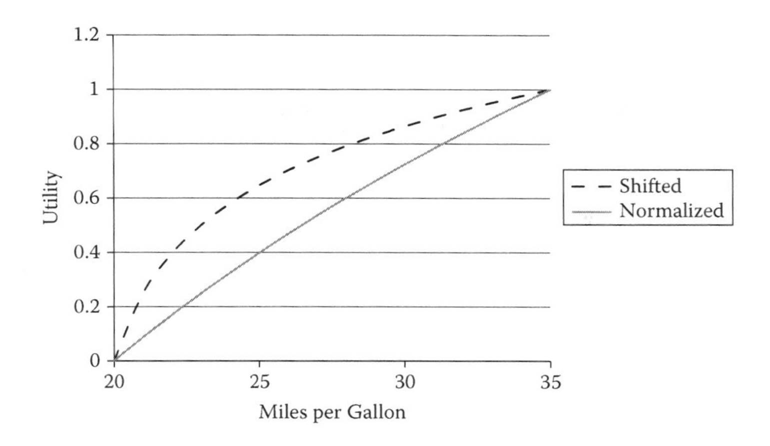

FIGURE 3.24
Shifting the utility function preserves the shape, while normalizing by subtracting the lowest utility value (Equation 3.21) flattens the curve.

on its domain. We do not have a lot of freedom with the exponential function because it uses only one parameter. It still makes sense to assess x for different values of probabilities and outcomes because then one can find the best-fit value of R. We will later show this method with a solved example. If different assessments give drastically different values of R, a different functional form should be chosen. Polynomials provide a lot of flexibility in assessing different preference structures at different points in the utility function, but they should be used only as a last resort because of instability issues.

Before we proceed, we draw the reader's attention to some finer issues to be considered while fitting utility functions. Notice that if $x = 20$ or 35, Equation 3.19 makes R equal to infinity. Clearly, the decision maker must provide a value *strictly between* 20 and 35 since there is a nonzero probability assigned to each of these extreme outcomes in the lottery. Another important observation to make is that to force the utility function to zero when $x = 20$, we shifted the value of the variables to the right. Other ways to accomplish the normalization are also possible; for example, see Equation 3.21. It will give the same result for the exponential function* but may be counterproductive for other functional forms. Figure 3.24 shows the comparison of normalizing the utility function using the two approaches when a logarithmic utility function is used. One can see that when one normalizes using Equation 3.21, the utility function is linearized. As a result, the ability of the utility function to correctly model the risk attitude of the decision maker can be compromised.

* We leave it for the reader to verify this fact. This property of the exponential utility function stems from the constant absolute risk aversion associated with it.

A fair amount of judgment therefore goes into the choice of utility function and how the different parameters are defined.

$$U_{0-1}(M)=\frac{U(M)-U(20)}{U(35)-U(20)} \quad (3.21)$$

3.8.3 Using Probabilities

The certainty equivalent method can be modified by asking the decision maker to provide preference probability values instead of the certainty equivalent. The advantages of doing this are threefold:

1. The decision maker is forced to think about probabilities.
2. The utility values are directly available, and they are equal to the indifference probability provided by the decision maker.
3. The utility function is by definition normalized between 0 and 1, since the value of probability is.

Let us revisit our fuel economy example. The certainty equivalent value is given to be 22 mpg. The decision maker is required to provide a value for the probability, p, of getting the best option (35 mpg), which automatically determines the probability of getting the worst option (Figure 3.25). For this method to work properly, the lottery must be a binary lottery involving the best and worst options in the range of negotiability. The CE values are changed and different values of the preference probabilities assessed. Notice that this method also implements the equivalence rule.

Since the utility associated with the best option is equal to 1 and that associated with the worst option is zero (without loss of generality), the preference probability provided for a sure value is equal to its utility. This can be proven easily. Let x be the sure value. For the decision maker to be indifferent between the lottery and the sure amount, we have

$$U(x)=pU(best)+(1-p)U(worst) \quad (3.22)$$

Replacing $U(best)$ with 1 and $U(worst)$ with 0, we get

$$U(x)=p \quad (3.23)$$

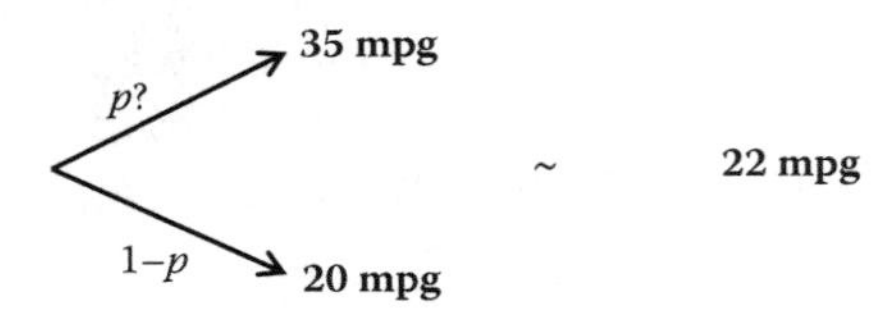

FIGURE 3.25
Assessing utility function using the indifference probability method.

which proves that the probability associated with the best option in the lottery gives the utility value of the sure option. The method is repeated with different values of x and the probabilities (which are also utility values) are recorded. A best-fit curve is then passed over the points assessed, which gives us the utility function. Note that the utility function is normalized between 0 and 1 by definition. We will show how this method is applied using a worked example.

Example 3.4

A decision maker wants to assess his utility function for money. He is actively considering prospects between \$3,500 and \$25,000. He is given lotteries to assess the preference probabilities associated with the intermediate amounts as given in Table 3.4. Find the best-fit exponential utility function. Also find the value of the utility function for \$10,000.

SOLUTION

Recall that if the best and worst options are given utility values of 1 and 0, respectively, the preference probabilities associated with the intermediate values are also the corresponding utilities. The problem can be solved using any optimizer or by using trial and error. The functional form chosen is

$$U(R,x) = \frac{1 - e^{-\frac{x - x_{min}}{R}}}{1 - e^{-\frac{x_{max} - x_{min}}{R}}} \tag{3.24}$$

The argument R is added on the left-hand side because it is the argument of optimization; usually if the utility function is known, it is a constant and is dropped. Notice closely that the exponential function is zero when the attribute is at the minimum acceptable level. Similarly, the denominator guarantees that the utility function is normalized to 1 at the maximum achievable level of the attribute. Although linear transformations on the utility function do not affect the decisions made, for proper fitting, it is a good idea to always use a properly normalized utility function.

TABLE 3.4

Preference Probabilities to Assess the Utility Function

Money (\$)	Preference Probability
3,500	0
7,000	0.3
14,000	0.7
18,000	0.85
20,000	0.9
25,000	1

We solve the following optimization problem, which looks for the value of R that minimizes the squared error along the utility curve, from the values in the table:

$$R = \arg\min \sum_{i=1}^{6} \left(U(R, x_i) - p(x_i)\right)^2 \quad R \in \Re^+ \tag{3.25}$$

where $p(x_i)$ is the preference probability from the table. The best-fit solution using a commercially available optimizer is found to be $R = 11{,}867.46$.

Therefore, the utility function is given by

$$U(x) = 1.195 - 1.195e^{-\frac{x-3500}{11867.46}}$$

Table 3.5 shows the values of the utility function from Table 3.4 against those predicted by the fitted utility function showing a close match. Using the best-fit utility function, the utility of \$10,000 is equal to 0.504.

3.8.4 McCord and de Neufville Procedure

McCord and de Neufville (1986) argued that the certainty equivalent method and the related probability method are flawed because of the certainty effect. Decision makers generally are biased toward certain outcomes and, as a result, will indicate extreme risk averseness in their utility functions. They proposed that for utility function assessment, a decision maker should be asked for indifference points between two lotteries, hence removing the certainty effect. Clearly the procedure requires extra effort from both the decision maker and the facilitator. The decision maker has to think harder to truly find the indifference point, while the facilitator has to potentially solve many equations to assess the utility function. The benefit, however, outweighs the effort required for most decision problems. Figure 3.26 shows the example lotteries to be compared.

Notice that there are many ways to introduce unknowns to look for indifference points between the two lotteries. The decision maker can be asked to

TABLE 3.5

Comparison of the Utility Values from the Assessment and Those from the Best-Fit Function

Money (\$), x	$p(x) = U(x)$	$U_{\text{fitted}}(x)$
3,500	0	0
7,000	0.3	0.305
14,000	0.7	0.702
18,000	0.85	0.843
20,000	0.9	0.898
25,000	1	1

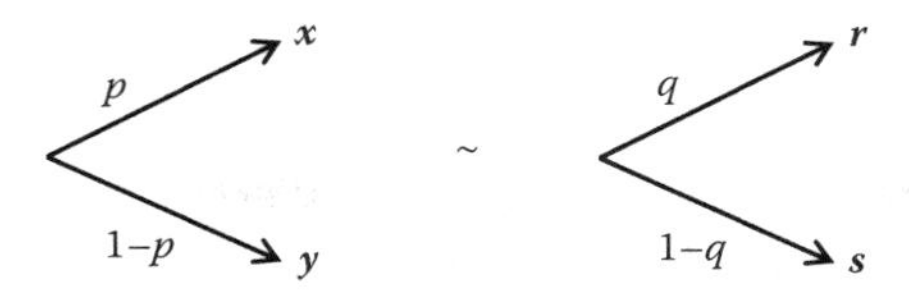

FIGURE 3.26
The McCord–de Neufville procedure: comparing two lotteries removes the certainty effect.

provide the probabilities p and q when everything else is fixed. Or he or she may be asked to provide the values of x, y, r, or s. In each case two lotteries are compared so the certainty effect is eliminated. To find the parameters for the utility function, the facilitator solves the following equation:

$$pU(x) + (1 - p)U(y) = qU(r) + (1 - q)U(s) \tag{3.26}$$

We will show how this method is applied using a worked example.

Example 3.5

Emily is indifferent between the following two lotteries (Figure 3.27). Assuming that she prefers more money to less, determine her utility function.

SOLUTION

We assume a utility function of the form $U(x) = a \log(bx + 1)$. Since the largest amount in the lotteries is \$100 and the smallest amount is \$0, we assign them utilities of 1 and 0, respectively. Therefore:

$$a \log(100b + 1) = 1$$

We also have from the lotteries:

$$0.4a \log(70b + 1) + 0.6a \log(10b + 1) = 0.8a \log(100b + 1) + 0.2a \log(0 + 1)$$

which gives us

$$\log((70b + 1)^{0.4}) + \log((10b + 1)^{0.6}) = \log((100b + 1)^{0.8}) + 0$$

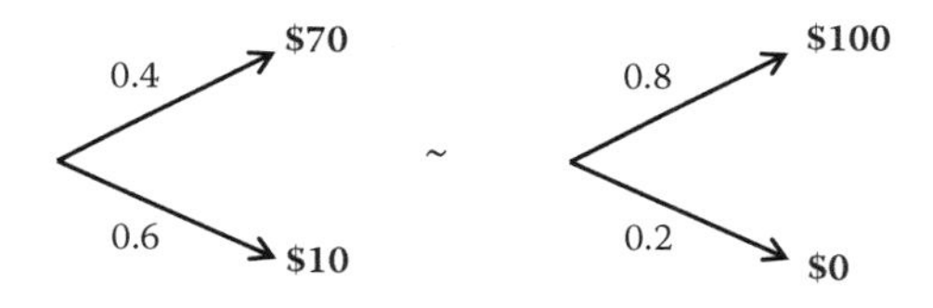

FIGURE 3.27
Assessing utility functions by comparing lotteries (Example 3.5).

The above equation can be solved numerically to yield $b = 20$, which makes a equal to 0.3029. The utility function is given by

$$U(x) = 0.3029 \log(20x + 1)$$

Note that the utility function we chose is not unique, and we could have chosen a different functional form just as easily. The basic properties we want to be able to reflect are ordering (more is better, or the other way around) and risk attitude. Depending on the scope of the decision, one can try different functional forms to get the best utility function to reflect the preferences.

3.9 Stochastic and Deterministic Dominance

One final topic we will discuss, before we move to multiple attributes, is that of dominance. Many times decisions are made easier if dominance exists between different options. Dominance is defined as the existence of a way to determine that a particular course of action will be better than another without having to perform all the calculations. Consider a simple scenario where we have to choose between two options, A and B. Option A pays a beta-distributed amount between \$100 and \$200, with mode at \$150 (Figure 3.28). Similarly, option B pays a beta-distributed amount between \$250 and \$350, with mode at \$300. Clearly, even the worst realization from B will be better than the best realization from A. As a result, any decision maker, regardless of his or her risk attitude, will prefer option B to A. We do not need to assess the utility function or find its expectation to make this determination. We define this situation as deterministic dominance. Therefore, option B deterministically dominates option A because even its worst realization is equal to or better than the best realization of A.

Consider now another option, C, that pays a beta-distributed amount between \$200 and \$300, with mode at \$250. Clearly, deterministic dominance does not exist between B and C (C does deterministically dominate A, though). However, we notice that for every fixed amount, the probability of C providing it or more (1 – CDF) is less than that for B. This is better represented using the CDFs of the payoff from the options. Notice that the

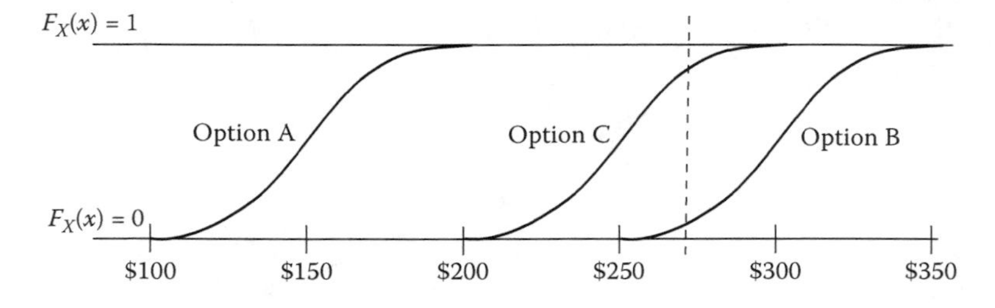

FIGURE 3.28
Stochastic and deterministic dominance between different options. Options B and C deterministically dominate option A. Option B stochastically dominates option C.

intersection of the vertical line at approximately \$270 with the CDFs shows that option B will provide \$270 or more with a higher probability than would C. This is true for all values on the x-axis. In this case we say that B *stochastically dominates* C. A decision maker will always prefer B over C. Again, we do not need to assess utility functions or calculate expectations.

Higher-order dominance conditions are also sometimes defined, exploiting further information about the CDFs. Some of them hold, for example, under the condition that a decision maker is risk-averse. In our opinion, the effort required in determining higher-order dominance is not always justified. However, we encourage readers to refer to publications that discuss higher-order dominance conditions, if they are so interested.

3.10 Multiattribute Decision Making

Engineering decisions almost always involve multiple attributes. Individually for each of these attributes, the preference structure is generally well formed in the decision maker's mind, but as a combination, they pose a challenge. For example, consider a wireless router. One usually prefers less cost to more, higher supported bandwidth to less, longer range to less, and so on. The question that is harder to answer is whether you would pay more to increase bandwidth by a certain amount if it concurrently decreases range by 10 m. There are some specific questions that arise when considering multiple attributes:

1. Do the five rules of actional thought hold in the case of multiattribute problems?
2. Can single-attribute preferences be translated into multiattribute preferences?
3. What tools are available to encode multiattribute preferences?

The five rules of actional thought apply to multiattribute problems as well if the decision maker can rank combinations of attributes. In other words, when applying order rule the decision maker should be able to say that one combination of attributes is better, worse, or equally desirable as another combination *and* the transitivity property holds. Assuming that such ranking is provided by a function U, defined over attribute vector $\mathbf{y} = \{y_1, y_2, \ldots, y_3\}^T$, the decision maker should be able to say (for two-attribute vectors $\mathbf{y}_1$ and $\mathbf{y}_2$):

$$U(\mathbf{y}_1) > U(\mathbf{y}_2) \tag{3.27}$$

or

$$U(\mathbf{y}_1) < U(\mathbf{y}_2) \tag{3.28}$$

or

$$U(\mathbf{y}_1) = U(\mathbf{y}_2) \tag{3.29}$$

Furthermore, transitivity should hold

$$U(\mathbf{y}_1) > U(\mathbf{y}_2) \wedge U(\mathbf{y}_2) > U(\mathbf{y}_3) \Rightarrow U(\mathbf{y}_1) > U(\mathbf{y}_3) \tag{3.30}$$

If the above conditions hold (order rule), then the five rules of actional thought apply readily. It is easy to see how the other four rules follow immediately.

The second question we want to answer is whether single-attribute functions can be translated to multiattribute preference. Under restrictive conditions, they can be. The independence conditions, as they are called, are discussed later in the chapter. Even when the independence conditions are not satisfied, there are ways to assess the utility function directly. The rest of the chapter is devoted to how multiattribute utility functions can be assessed and used in decision making (which answers the third question). Here we will spend some time understanding trade-offs and how to represent them graphically using Pareto fronts and mathematically using multiattribute value and utility functions.

3.10.1 Trade-Offs

In our router example we hinted at trade-offs. Trade-offs are inevitable in any real-life decision-making situation, and engineering design is no exception. There are two contexts in which the term *trade-off* is used in engineering:

1. **Trade-off behavior:** The willingness and degree to which a decision maker is willing to give up performance in an attribute to get improvement in another.
2. **Unavoidable trade-offs:** The trade-offs that are inherent to a decision problem because improvement in an attribute invariably comes with a worsening of another.

Engineering design therefore can also be looked at as reconciling the trade-off behavior of the decision maker with the unavoidable trade-offs that the design problem presents.

3.10.1.1 Trade-Off Behavior

When faced with decision problems involving multiple attributes, starting with a baseline option, every change in one attribute must be compensated exactly by another, for the decision maker to remain indifferent. If it is not true, then the decision maker prefers one option over another. An option in this discussion is the attribute vector, $\mathbf{y} = \{y_1, y_2, \ldots, y_3\}^T$, where the y_i's are the attribute levels, each a function of the design decision variable vector $\mathbf{x}$. For consistency, and without lack of generality, we assume minimization of these attributes.* Figure 3.29 shows attribute space for two variables y_1 and

* All maximizations can be converted to minimization by a suitable transformation, e.g., adding a negative sign.

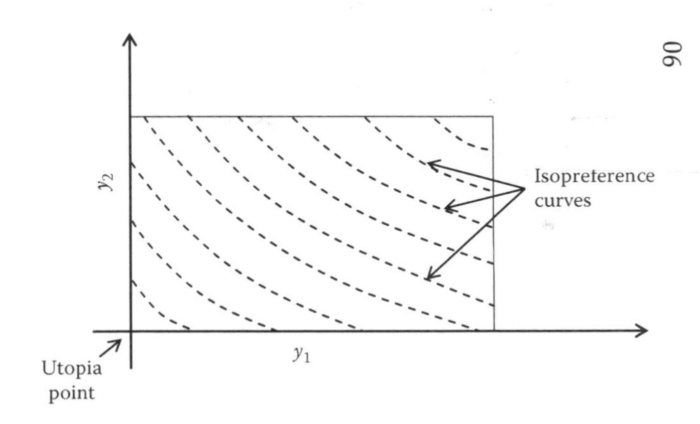

FIGURE 3.29
Trade-off behavior represented by the isopreference curves in a two-attribute problem.

y_2. The ranges on the horizontal and vertical axes are the ranges of negotiability for the two attributes. Clearly the origin is the best desirable option (utopia point), and the point on the top right is the worst acceptable option. The figure shows curved lines called *isopreference* curves or lines. Any movement along these curves changes values of the attributes. The change, however, is such that the satisfaction level (utility) of the decision maker does not change—hence the term *isopreference.* For any point in the space any movement into the bottom-left quadrant respective to the point improves the utility, while any movement into the top-right quadrant decreases the utility, as shown in Figure 3.30.

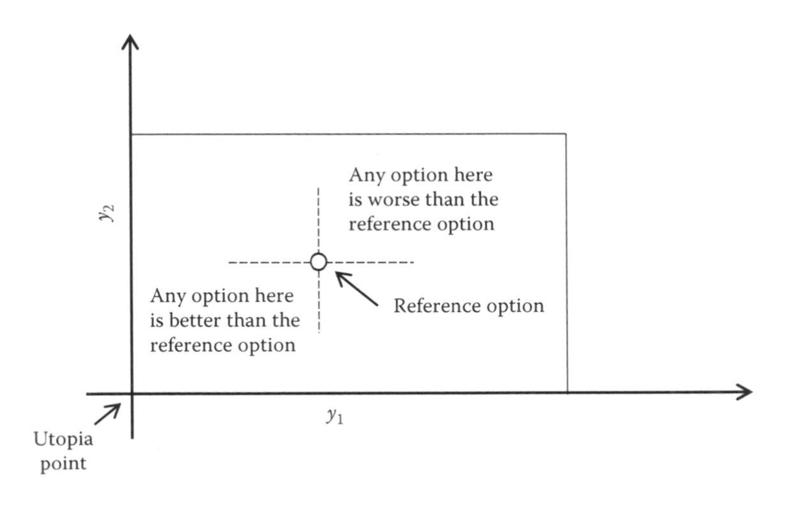

FIGURE 3.30
Dominance in a two-attribute problem.

The trade-off behavior of the decision maker is embedded in his or her utility function. A way to look at an isopreference curve is to realize that the utility function stays constant along that curve.

3.10.2 A Pareto Front

A Pareto front is a widely utilized tool for visually depicting unavoidable trade-offs between different attributes a decision maker is actively considering. The trade-off is a constraint imposed by factors beyond the control of the decision maker; that is, there is no way to avoid this trade-off. If trade-offs were not necessitated by the decision problem, one could increase a favorable attribute indefinitely without having to consider the detrimental effects on other attributes. Our experience, and most possibly that of the reader, tells us that this is obviously not true.

Consider two attributes that we are trying to minimize, y_1 and y_2. An example Pareto front is shown in Figure 3.31. As before, the utopia point is the origin. A curve like the one shown can be called the true Pareto front for a particular design problem if it is not possible to find a solution that is to the left and below (simultaneously) of any point on the curve; that is, the curve provides the best possible trade-offs. If we follow the curve by moving left to right (thereby worsening y_1), we see an improvement in y_2. Similarly, improving y_1 by moving right to left worsens y_2. More generally, for n attri-

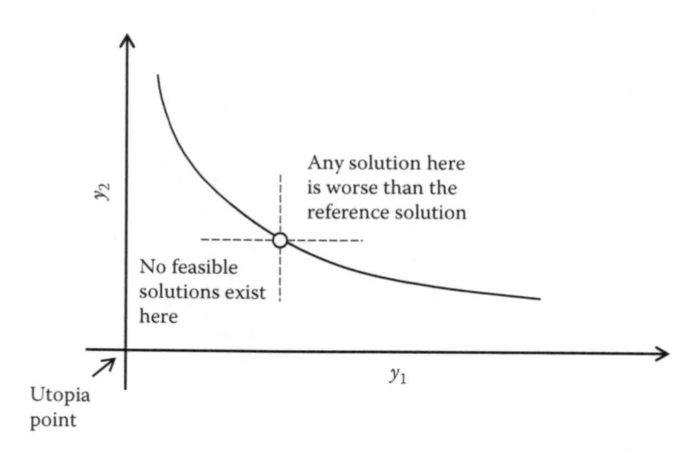

FIGURE 3.31
An example Pareto front for a two-attribute problem.

butes, a solution $\mathbf{x}^P$ belongs on the Pareto front if there is no other solution $\mathbf{x}^{P*}$, such that

$$y_i\left(\mathbf{x}^{P*}\right) \le y_i\left(\mathbf{x}^P\right) \quad \forall i \in \{1,..,n\} \text{ and } \exists j \in \{1,..,n\} \text{ such that } y_j\left(\mathbf{x}^{P*}\right) < y_j\left(\mathbf{x}^P\right) \quad (3.31)$$

A Pareto front can be defined on as many attributes as one desires; however, *finding* a Pareto front over many attributes and visually depicting it becomes a challenge. Except for some simple problems, generating a Pareto front is accomplished using multiobjective optimizers designed for that purpose. There are mainly two ways to do this: optimizing convex combination of attributes and nondominated sorting.

1. **Optimizing convex combination of attributes:** Let us say that we have the attributes $y_i(\mathbf{x})$ and the decision variable is the vector $\mathbf{x}$. Without loss of generality, we can assume that they are to be minimized.* A convex combination of the attributes is given by

$$y(\mathbf{x}) = \sum_i \alpha_i y_i(\mathbf{x}) \quad (3.32)$$

   where $\sum_i \alpha_i = 1$ and $\alpha_i \in [0,1]\ \forall i$.

   The convex combination $y(\mathbf{x})$ is then minimized subject to the constraints of the problem, for a given combination of α_i's. The solution will give us one point in the attribute space that lies on the Pareto front. We repeat the optimization process for all the possible combinations of α_i's and plot the solutions in the attribute space. The set of the solutions forms a Pareto front. Ideally, we would want to implement all combinations of α_i's, but since optimization problems are generally computationally expensive, the coefficients are varied in a discrete way. Finer discretization can be done in the case where one wants higher resolution in the Pareto front.

2. **Nondominated sorting:** The method of nondominated sorting involves working with a set of solutions. Each solution has values of the attributes associated with it. Initially a set of solutions is randomly generated. This set can be seeded with solutions that have

* If an objective is to be maximized, we can define another function that is the negative of the attribute and minimize that function.

been shown to be historically correct or that provide a good building block. The solutions are then compared against each other as follows (n attributes).

A solution $\mathbf{x}_1$ dominates a solution $\mathbf{x}_2$ if

$$y_i(\mathbf{x}_1) \leq y_i(\mathbf{x}_2) \quad \forall i \in [1, n] \tag{3.33}$$

and

$$\exists j \in \{1, \ldots, n\} \quad \text{such that} \quad y_j(\mathbf{x}_1) < y_j(\mathbf{x}_2) \tag{3.34}$$

In words, the above relations state that a solution is said to dominate another if it is at least as good as the other solution in all the attributes and it is strictly better in at least one attribute. Dominance does not have to exist between a pair of solutions. For our example it is possible that the above relationships do not establish that $\mathbf{x}_1$ dominates $\mathbf{x}_2$, nor the equally possible case that $\mathbf{x}_2$ dominates $\mathbf{x}_1$. In such cases the solutions are mutually nondominated. The immediate question then is: Are there solutions for which we cannot find *any* feasible solution that dominates them? When such solutions are found, they fall on the Pareto front. A Pareto front is a set of solutions that are mutually nondominated and there are no feasible solutions that dominate any one of the points on the front.

Needless to say, finding the Pareto front using domination requires sampling all of the solutions, or if that is not possible, a vast number of them. Evolutionary algorithms are generally used because they work with a set of solutions. Using the operators specific to an algorithm, new and better solutions are found and tested for dominance. Once convergence criteria are met, one is left with an approximation of the Pareto front that serves the purpose for most cases. In Chapter 5, we will go over multiobjective optimization using evolutionary algorithms in detail.

3.10.3 Limitations of Pareto Fronts

Pareto fronts are very helpful when making multiattribute trade-off decisions. One crucial aspect of a decision situation for which they cannot account, however, is uncertainty. Pareto fronts only show trade-offs between deterministic choices. Since all design decision problems involve uncertainty, this proves to be a huge limitation. Pandey, Mourelatos, and Nikolaidis (2011) show that not only do Pareto fronts fail to capture trade-offs under uncertainty, but there is no way of representing this trade-off, that is, with any function of the attributes, unless the decision maker trades off

TABLE 3.6

Decision Making in the Presence of a Design Problem as well as Preference Assessment Uncertainty

	No PA Uncertainty	PA Uncertainty Present
No IDP uncertainty	Maximize multiattribute utility function using deterministic optimization	The DM selects a design from a deterministic Pareto front. An explicit multiattribute utility function is not required.
IDP uncertainty present	Maximize expected multiattribute utility function by controlling the statistics of the design variables	• If the DM trades off single-attribute utilities linearly, use a Pareto front with respect to the expectation of utilities. • Devise new method. Pandey et al. (2011) propose a method using certainty equivalent curves.

the single-attribute utilities linearly. Two kinds of uncertainties are defined in that work: inherent design problem (IDP) uncertainty, or uncertainty pertaining to the stochasticity in the decision variables, and preference assessment (PA) uncertainty, pertaining to lack of knowledge of the preferences of the decision maker. Arguably, both kinds of uncertainty are present in all design decision problems. Table 3.6 sums up the design decision-making methodology, particularly in light of this information.

3.10.4 Different Multiattribute Functions

If all we know about the decision maker's preferences is his or her ordering for each attribute, we has to resort to a Pareto front. The decision maker is then presented with these solutions on the Pareto front and chooses one solution or design. Recall that the five rules of actional thought imply that a utility function exists, regardless of whether the decision maker or the facilitator is aware of it. Eliciting and encoding this function has a lot of value in decision based design because finding the optimal decision can be automated, and the computational difficulty of finding the Pareto front is also mitigated. Multiattribute utility functions are supposed to capture three basic elements of the preferences of the decision maker:

1. The preference order for each attribute throughout the design space
2. The trade-off behavior of the decision maker over multiple attributes
3. The risk attitude of the decision maker over the design space

Clearly, acquiring multiattribute utility functions is harder than single-attribute functions. This is because of the *curse of dimensionality*. Each solution forms a tuple in the attribute space, and comparing it with another has to account for the above three elements. Also, to get the mathematical form of the multiattribute function, one has to pose many questions to the decision maker. Approximations and simplifying assumptions can sometimes help in acquiring the function. In the next few sections, we will go over various ways of accomplishing this task.

Some notes on nomenclature: In this section we will represent multiattribute utility functions with $U_M(.)$, which mathematically measures the desirability of an option involving multiple attributes. As with they are single-attribute functions, they are scaled between 0 and 1. A value of 0 implies that all the attributes are at the worst acceptable level, and a value of 1 implies that all the attributes are at the best possible level. When the function is written as $U_M(y_1,...,y_n)$, we talk about multiattribute utility as a function of raw attributes, while $U_M(U_1,...,U_n)$ refers to the multiattribute utility as a function of single-attribute utilities.* A preference order on the ith attribute, y_i, is defined by the sign of $\frac{\partial U_M(y_1,...y_n)}{\partial y_i}$, which simply tells us whether we prefer more of y_i over less, or the other way around, that is,

$$PO_{y_i} = sign\left(\frac{\partial U_M(y_1,...y_n)}{\partial y_i}\right) \tag{3.35}$$

A positive value of PO_{y_i} indicates that the decision maker prefers more of y_i to less, and a negative value indicates that the decision maker prefers less of y_i over more. Notice that we do not necessarily require the preference order to remain constant for all values of other attributes (even though it makes things easier, as we shall see later). In other words, the partial derivative value can potentially change based on the value of other attributes. Furthermore, the definition does not necessarily require differentiability; a definition based on finite differencing works equally well: $PO_{y_i} = sign\left(\frac{\Delta U_M(y_1,...y_n)}{\Delta y_i}\right)$. However, it is generally a good idea to use differentiable utility functions, as the derivative also tells us how quickly the desirability changes with respect to the attribute in question.

3.10.5 Independence Conditions for Combining Single-Attribute Functions into a Multiattribute Utility Function

When single-attribute utility functions are available, the multiattribute utility function can be determined using some function of these single-attribute utility functions. Some independence conditions, however, need to be satisfied. In this section we will discuss three such conditions.

* Note that individual utilities, U_i's, are functions of raw attributes themselves, but writing the multiattribute function with U_i's as arguments shows that the utility function is separable. This point will become clear later in the section.

3.10.5.1 Preferential Independence

If the preference order over one attribute does not change regardless of where the values of other attributes are fixed, it is preferentially independent of the other attributes, that is, y_i is preferentially independent of other attributes if in its domain (range of negotiability)

$$PO_{y_i} = \text{constant}, \quad \forall y_{j,fixed} \in dom(y_j),\ j \neq i \tag{3.36}$$

Furthermore, if the above relationship holds for all attributes in question, then the attributes are called *mutually preferentially independent*. Preferential independence is relatively easy to satisfy and easy to verify. For example, in most purchasing decisions cost is preferentially independent of other attributes; we prefer lower cost to more, regardless of where the other attributes are fixed. Similarly, we try to make designs cheaper and faster to manufacture. On the other hand, attributes that are seemingly intertwined in the decision maker's mind may not be preferentially independent. For example, the acceleration of a car and price of gasoline may not be preferentially independent. Higher acceleration usually comes with a lower fuel economy; therefore, when gasoline is cheap, a decision maker will prefer high acceleration, and similarly, when gasoline is expensive, he or she may prefer a lower-acceleration car. Notice that the decision maker could also be probably confusing correlation for causation (high acceleration is not caused by lower fuel economy; it could also be a result of a better, more efficient engine). If we use fuel economy as a surrogate for gasoline price, we find that it is preferentially independent of acceleration. For a fixed fuel economy, the decision maker prefers higher acceleration, and for a fixed acceleration, he or she prefers higher fuel economy. Careful redefining of attributes almost always removes preferential dependence.

While preferential independence can generally be ascertained without using numbers, it is a good idea to use them. One would, for example, set the fuel economy at 25 mpg and ask the decision maker if he or she prefers more acceleration to less *over the whole range of negotiability* for acceleration. Next, the fuel economy is set at some other value and the same question asked again. This process is repeated until the facilitator is certain that the preference order will not change regardless of where the fuel economy is fixed in its range of negotiability. Next, the process is repeated with the attributes switched. For example, one might fix the 0–60 acceleration at 6 seconds and ask the decision maker if he or she prefers lower fuel economy to higher. After repeating this process a few times, the facilitator (and the decision maker) can determine if the two attributes are mutually preferentially independent.

3.10.5.2 Utility Independence

An attribute is utility independent of others if the certainty equivalent of any lottery over the attribute has the same value regardless of where the values of the other attributes are fixed.

$$CE_{y_i} = \text{constant}, \quad \forall y_{j,fixed} \in dom(y_j),\ j \neq i \tag{3.37}$$

Consider the above decision situation with fuel economy and acceleration. Let us say that the decision maker is asked to value a car that provides a 0–60 acceleration of 6 seconds, but the fuel economy is normally distributed with mean 25 mpg and standard deviation 2 mpg. Assume a risk-averse decision maker who, let us say, values the fuel economy at 23 mpg. Now if we set the acceleration at any other value without changing the distribution of the fuel economy and the decision maker still values the fuel economy at 23 mpg, then fuel economy is utility independent of acceleration. Similarly, we test for utility independence of acceleration from fuel economy. We set the fuel economy fixed at a certain value and give the decision maker a probability distribution over acceleration. He or she is then asked how much he or she values acceleration ($CE_{\text{acceleration}}$). If this value of $CE_{\text{acceleration}}$ does not change regardless of where we fix fuel economy, acceleration is also utility independent of fuel economy. The two conditions imply that the attributes are mutually utility independent. In case of many attributes, every attribute should be tested for utility independence. While it sounds tedious, usually testing at a few points within the ranges of negotiability is enough to be confident that attributes are utility independent.

It is generally a good idea to test for preferential independence first; if it is not satisfied, utility independence will most likely not be satisfied either. Also, utility independence should be checked for all attributes to determine if they are mutually utility independent. If utility independence is not satisfied, then one might try to redefine attributes so that they become independent. Utility independence is a generally harder condition to satisfy than preferential independence.

3.10.5.3 Additive Independence

Additive independence applies to cases where uncertain choices over one attribute are not affected by uncertain choices over another. As such, it is the hardest of the independence conditions to satisfy. The formulation is as shown in Figure 3.32. Let the two attributes of acceleration and fuel economy be denoted by a and f, respectively, while the indices represent different constant values. There are two 50-50 lotteries between attribute combinations. Additive independence requires that the decision maker should stay indifferent if the attributes' values are switched, as shown in Figure 3.32. Additive

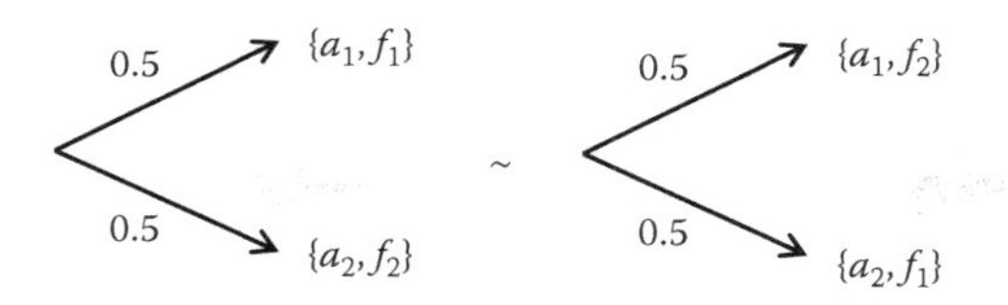

FIGURE 3.32
Test for additive independence.

independence should exist for all the attributes for them to be mutually additively independent. Additive independence implies utility independence.

3.10.5.4 Independence Conditions, Correlation, and Ranges of Negotiability

Independence conditions determine whether the attributes are independent in the minds of the decision maker. This independence has nothing to do with statistical correlation that almost always exists in design problems. Many times naïve decision analysts confuse the two. Two attributes could be perfectly correlated but still independent in a decision maker's mind. In fact, if the attributes were not correlated, we could improve one without affecting the other, indefinitely. This is definitely not the case, as we know from our experience.

Recall that we stressed the importance of determining the range of negotiability for attributes. Proper ranges of negotiability are even more critical in multiple-attribute problems. This is because the independence conditions are required to hold only within the ranges of negotiability of the attributes. When the ranges of negotiability are arbitrarily large, even preferential independence becomes hard to satisfy. It is therefore important to have well-defined ranges for all the attributes. One note of caution, though: ranges of negotiability should not be arbitrarily reduced to satisfy the independence conditions. Value function-based formulations must then be used, as we will discuss in later sections.

3.10.5.5 Why Do We Test for Independence Conditions? The Multiplicative Form

The independence conditions ensure some fundamental properties of a decision situation. Single-attribute utility functions can be directly used to determine multiattribute utility. For example, if the attributes are preferential and utility independent, one can use the multiplicative form given by

$$U_M(U_1,\ldots,U_n) = \frac{1}{K}\left(\prod_{i=1}^{n}(Kk_iU_i + 1) - 1\right) \tag{3.38}$$

The single-attribute utility functions are assumed known. The (strictly positive) scaling constants k_i show the relative *disinclination** of the decision maker to trade off the concerned attribute for improvement in other attributes. While constrained to lie between 0 and 1, the higher the value of the scaling constant, the less likely the decision maker is to accept worsening of the attribute. The normalizing parameter, K, guarantees that U_M is constrained between 0 and 1. Its value is completely determined by the scaling constants k_i. To calculate K, the U_i's are set equal to 1, and so is U_M which results (for n attributes) in a polynomial equation of order $n - 1$. The equation is then solved analytically or numerically to find K. If k_i's add up to more than 1, the value of K is between –1 and 0. If k_i's add up to less than 1, then K is positive. This test is important because the polynomial results in multiple roots, not all of which are usable. Keeney and Raiffa (1994) show the reasoning behind this choice.

If the scaling constants add up to more than 1, the attributes act as substitutes; that is, satisfaction from one attribute can compensate for that from another. The cross-utility terms (product terms in the expanded form of Equation 3.38) actually lower the contributions from individual attributes. Similarly, when the scaling constants add up to less than 1, the cross-utility terms contribute positively. In such cases the attributes are called complements.

When only two attributes are present, the multiplicative formulation breaks down into the multilinear form given by

$$U_M(U_1,U_2) = k_1U_1 + k_2U_2 + (1 - k_1 - k_2)U_1U_1 \tag{3.39}$$

Notice that we do not need to explicitly determine K in the two-attribute case.

When all three independence conditions (preferential, utility, and additive) are satisfied, we can use a weighted linear sum of the single-attribute utilities, that is,

$$U_M(U_1,\ldots,U_n) = \sum_{i=1}^{n} k_iU_i \qquad \sum_{i=1}^{n} k_i = 1 \tag{3.40}$$

Realize that the scaling constants do not necessarily have to add up to 1 (for the case in Equation 3.40); it is done because we conventionally want the multiattribute utility function to be normalized between 0 and 1. Notice that there are no cross-utility terms; each attribute contributes somewhat independently to the multiattribute utility and hence additive independence. It is unfortunate that the weighted sum is used indiscriminately without realizing what strict conditions need to be satisfied for its applicability. The converse is also true in that when the scaling constants add up to 1, the

* Sometimes the k's are said to signify the importance of an attribute. This is not correct. An attribute with a low k value can be extremely important, while an attribute with a large k can be unimportant.

attributes are additively independent. One should revisit assessment of scaling constants in such cases.

3.10.6 Assessing Multiattribute Functions

When independence conditions are satisfied and single attribute functions are known, assessing multiattribute functions is reduced to finding the values of the scaling parameters. Standard fitting techniques generally apply. The lottery method is commonly used where the decision maker is asked to provide an indifference probability for a lottery versus a fixed alternative. The lottery includes an alternative with all the attributes at the best possible level and an alternative with all the attributes at the worst possible level. The fixed alternative has all the attributes at the worst acceptable level, except the attribute in question, which is at the best possible level. Figure 3.33 provides the formulation.

The validity of the above assessment method can be readily proven. The indifference probability satisfies the following equation:

$$U_M\left(y_{1,worst},\ldots,y_{i,best},\ldots,y_{n,worst}\right)=pU_M\left(y_{1,best},\ldots,y_{n,best}\right)+\left(1-p\right)U_M\left(y_{1,worst},\ldots,y_{n,worst}\right) \tag{3.41}$$

By convention, the multiattribute utility is 1 when all attributes are at the best possible level, that is, $U_M(y_{1,best},\ldots,y_{n,best})=1$, and the multiattribute utility is 0 when all the attributes are at their worst possible level, that is, $U_M(y_{1,worst},\ldots,y_{n,worst})=0$. This gives us

$$U_M\left(y_{1,worst},\ldots,y_{i,best},\ldots,y_{n,worst}\right)=p \tag{3.42}$$

Now if we set all the attributes at their worst possible level, their single-attribute utility is 0. We do this for all attributes except the attribute in question, which is at the best possible level, and thus its utility is 1; we get from the most general form of the multiplicative form (Equation 3.38)

$$U_M\left(0,\ldots,1,\ldots,0\right)=k_i \tag{3.43}$$

Comparing the two equations, we get $k_i=p$. In other words, the scaling constant associated with an attribute in the multiplicative form of utility

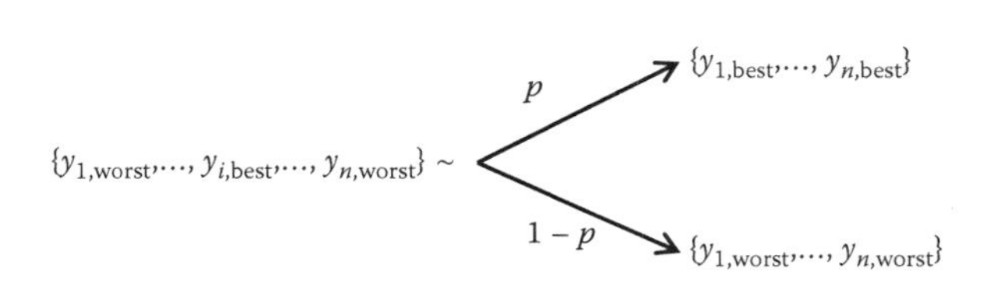

FIGURE 3.33
Assessing scaling constants in multiattribute problems.

function is equal to the indifference probability of a lottery between attributes at the best possible level and those at the worst acceptable level, with a fixed alternative that has only the attribute in question at the best possible level, while others are at their worst possible level. The indifference probabilities need to be found for all the attributes to get the complete multiattribute utility function. This discussion also shows that when only one attribute is at the best possible level while others are at their worst possible level, the multiattribute utility is equal to the scaling constant associated with the attribute.

3.10.7 Utility over Value Functions

When the independence conditions are not satisfied, one cannot use the multiplicative function or any of the special cases of it we have described in the previous section. In such cases we define a value function over the attributes and use a single-attribute utility function over the value function. Mathematically, the attributes y_i are lumped together in a value function, $V(y_i,...,y_n)$, which can rank deterministic options. The utility function $U_M(V)$ then incorporates the decision-making preferences under uncertainty to give the value of multiattribute utility. This method is more general because it does not require independence conditions to be satisfied, but requires assessment of the value function at various points in the ranges of negotiability.

Recall from earlier in the chapter that a value function helps rank alternatives when no uncertainty is present. A multiattribute value function must look at *combinations* of attributes and be able to rank them. The process is similar to any curve-fitting technique. We illustrate this with the following example. Consider Bill, who is trying to buy an electric drill. The only two attributes he is considering are rpm (s) and cost (c). The first thing we want to do is find the ranges of negotiability for the two attributes. Let us say that the ranges of negotiability for the attributes are $\{s_{\min}, s_{\max}\} = \{300, 3000\}$ for rpm and $\{c_{\min}, c_{\max}\} = \{\$20, \$100\}$ for cost. Bill prefers high rpm to low and low cost to high. While many functional forms are possible, we define the value function as follows:

$$V(s,c) = \frac{(s-s')^a}{(c-c')^b} \tag{3.44}$$

There are four unknown parameters in the above equation: s', c', a, and b. We can assess the values of s' and c' by asking him more questions, but for the sake of this example, let us assume values of 200 rpm and \$10, respectively. Also, b is assumed to be 1. Substituting these values into the equation above, we get

$$V(s,c) = \frac{(s-200)^a}{(c-10)} \tag{3.45}$$

Now Bill is asked some questions to determine his trade-off behavior over deterministic combinations of the two attributes. For example, consider two drills with the attributes at {500 rpm, $30} and {1,500 rpm, *c*?}. Bill is asked what value of the cost of the second drill will make him indifferent between the two drills. Clearly, the value should be greater than $30 because the second drill has higher rpm. Let us say that Bill says he is willing to pay $50 for the second drill. We can now write the indifference equation:

$$\frac{(500-200)^a}{(30-10)} = \frac{(1500-200)^a}{(50-10)}$$

$$\frac{(300)^a}{20} = \frac{(1300)^a}{40}$$

or

$$4.333^a = 2$$

Taking logarithms of both sides we get

$$a \log 4.333 = \log 2$$
$$a = 0.4727$$

Therefore, the value function becomes

$$V(s,c) = \frac{(s-200)^{0.4727}}{(c-10)} \tag{3.46}$$

Now, we need to find the utility function over this value function. Notice that we want the utility to be 0 when the attributes are at their worst acceptable level. One straightforward way of doing this is to assume the exponential functional form and make the value function equal 0 at the least acceptable level of the attributes. We modify the above function as

$$V'(s,c) = \frac{(s-200)^{0.4727}}{(c-10)} - V(300,100) \tag{3.47}$$

or

$$V'(s,c) = \frac{(s-200)^{0.4727}}{(c-10)} - 0.098$$

Another normalization we want to do is to make sure that the utility is equal to 1 when the attributes are at their best possible level. We

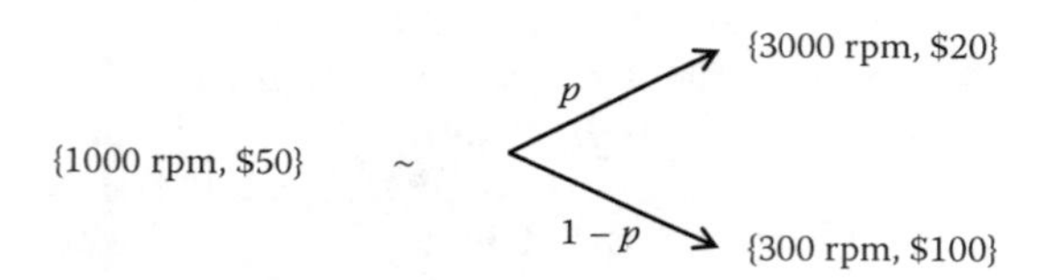

FIGURE 3.34
Assessing utility function on top of value functions (Bill's drill-buying problem).

therefore define the following functional form of the utility function over the value function:

$$U(s,c) = \frac{1-\exp(-V'(s,c)/R)}{1-\exp(-V'(3000,20)/R)} \tag{3.48}$$

We are now down to assessing a single-attribute utility function, which we have already learned how to do in this chapter. Bill is presented with a decision where he chooses the indifference probability of a lottery between the best and worst outcomes against a deterministic alternative with intermediate values of attributes. Figure 3.34 shows an example. Let us say that his indifference probability is 0.7. We can solve the following equation for R:

$$\frac{1-\exp(-V'(1000,50)/R)}{1-\exp(-V'(3000,20)/R)} = 0.7 \tag{3.49}$$

which gives the value of R as 0.407. The utility function is therefore

$$U_M(s,c) = 1-\exp\left(-\frac{(s-200)^{0.4727}/(c-10)-0.098}{0.407}\right) \tag{3.50}$$

The reader can verify that this function is consistent with all the answers provided by Bill.

3.10.8 Attribute Dominance Utility and Multiattribute Utility Copulas

Most of the treatment considered so far in this book is what can be called classical decision analysis. Decision analysis is an active area of research where new methods of defining and eliciting utility functions are constantly being developed. A recently proposed approach uses an analogy between probability distributions and utility functions (Abbas and Howard, 2005). A joint distribution is similar to a multiattribute utility function in that it is increasing in individual arguments and scaled between 0 and 1. This method leads to an attribute dominance utility because it is equal to 0 when any of the attributes is equal to its lowest acceptable level. Under these conditions, the utility function resembles a cumulative probability distribution function (except the property of cumulative probability distribution functions that they are n-increasing, i.e., have a positive mixed derivative).

The analogy has many intuitive advantages. It brings a lot of results in probability theory into our arsenal. Abbas and Howard (2005) show how the attribute dominance formulation allows for notions such as utility inference along the lines of Bayes' rule. As mentioned earlier, in the case of multiple attributes, utility functions do not satisfy the condition of a positive mixed derivative, whereas probability distributions, by definition, do. Abbas (2009) shows how this issue can be resolved using multiattribute utility copulas.

3.11 Biases in Decision Making

We conclude this chapter with a discussion on how our personal biases affect our decision making considerably. Tversky and Kahneman (1974) identify some common biases that play a role in decision making, particularly in assessment of probabilities. We follow the discussion from Nikolaidis et al. (2011).

1. **Representativeness:** Representativeness bias occurs when we believe that an outcome is likely to occur just because we observed a set of events historically associated with the outcome. Representativeness results from incorrect associations. A doctor who is used to seeing many malaria cases in his geographical area might be more likely to make malaria diagnoses if he sees related symptoms in patients, even though the symptoms could be associated with many other diseases.
2. **Availability:** Availability bias while assessing probabilities is seen when a person can recall from his or her recent memory an incident similar to the one in question. For example, an oil company that has had the misfortune of an explosion in its offshore oil rig will overestimate the probability of it happening again, when commissioning a new rig.
3. **Anchoring:** Anchoring happens in probability assessment as well as when considering new ideas. It refers to the human mind getting fixated on an idea or a number because the person has heard it recently. For example, an experienced designer will have a hard time designing a radically new suspension for a vehicle because he or she is so used to the idea of a spring-damper system. Similarly, while assessing the probability of an outcome, as soon as a probability number is mentioned, most decision makers will fixate on that number and provide values around it.
4. **Overconfidence:** Overconfidence bias affects us all some time or another. Overconfidence is the unwillingness to consider others' opinions or

believe data because we overestimate our likelihood of being right. This bias is so ingrained in most decision makers' thinking that it is hard to even provide true confidence intervals even when there is no incentive in being correct! Real-life examples of this bias can be easily seen by asking people 95% confidence intervals for numbers associated with commonly discussed items, for example, height of Mount Everest, age of the earth, distance of Venus from the earth, number of red blood cells in the human body, number of suicide deaths in a year.

5. **Motivational:** Many times when reporting probabilities of uncertain events decision makers inflate or decrease the values based on what suits them the most. For example, while trying to sell used machinery, people underestimate the failure probability of the machinery.

Example 3.6

A mechanical engineer is designing the intake manifold for an internal combustion engine. He has identified three critical dimensions, d_1, d_2, and d_3, in centimeters, along different points on the manifold that he thinks are critical to the two attributes with which he is concerned: fuel economy (η in miles per gallon) and 0 to 60 mph acceleration time (a in seconds). He has found a simplistic model to represent this relationship after lengthy simulations of the engine for different values of d_1, d_2, and d_3.

$$\eta = 0.8d_1d_2 - \sqrt{d_3}$$

$$a = \frac{3d_1d_2}{d_3}$$

He provides the indifference points shown in the following table between the two attributes. Fit a suitable multiattribute value function and also find the value of d_1, d_2, and d_3, all constrained to be less than 10 cm, that maximizes the value function.

Attributes Tuples for Example 3.6

Attribute Tuple 1	Attribute Tuple 2
{30, 12}	{35, 14}
{25, 10}	{22, 9}
{30, 8}	{35, 10}
{20, 7}	{35, 11}

Note: Options in the two columns of the same row are equally desirable.

SOLUTION

We assume a value function of the form

$$V(\eta, a) = \eta a^{-b} \tag{3.51}$$

where b is strictly positive. The above formulation increases the value in efficiency and decreases the time to 0–60 acceleration, as expected. The best-fit function is found for $b = 1.2133$ that minimizes the error between the value function for the attribute tuples in the two columns (see the following table). Using a commercial optimizer, the values of d_1, d_2, and d_3 are found to be 2.44, 8.98, and 9.97 cm, respectively, that maximize the value function.

Attribute Tuples for Example 3.6 and the Corresponding Value Function Values for $b = 1.2133$

Attribute Tuple 1	Value Function	Attribute Tuple 2	Value Function
{30, 12}	1.471	{35, 14}	1.424
{25, 10}	1.529	{22, 9}	1.529
{30, 8}	2.407	{35, 10}	2.142
{20, 7}	1.887	{35, 11}	1.907

Example 3.7

The efficiency of a windmill tested at different wind speeds is given in the following table. The intermediate efficiencies can be approximated using a polynomial. Two locations are being considered, location A and location B. The wind speed is normally distributed with mean 10 m/s and standard deviation 2 m/s in location A. While in location B the distribution is normal, with mean 22 m/s and standard deviation 5 m/s. Which location should be chosen if the utility function of the company installing the windmill is a function of efficiency as

$$U(\eta) = 1 - e^{-\frac{\eta}{10}} \tag{3.52}$$

Wind Speed and Efficiency Table for a Windmill

Wind Speed (m/s)	Efficiency (%)
1 and lower	0
5	5
10	25
15	26
20	20
35	5
45 and higher	0

SOLUTION

The best-fit polynomial of order three is found using a commercial solver. The equation is

$$\eta = 0.0023x^3 - 0.2022x^2 + 4.5857x - 6.9205 \tag{3.53}$$

where η is efficiency and x is the wind speed. One thousand realizations of wind speed were generated for each location, and the corresponding values of the efficiency were calculated using Equation 3.53. For each realization the utility function value was calculated using the efficiency value. Expected utility was then found by averaging the utility values for each location. The expected utility values are 0.866 and 0.838 for the two locations, and therefore location A should be chosen.

Problems and Exercises

1. What is a good decision? Give an example of a good decision that you made in your life that resulted in a bad outcome. Also give an example of a bad decision that resulted in a good outcome.
2. How does uncertainty affect our decision making?
3. What are axioms? Why do we need them?
4. What are the five rules in decision analysis? Write a short description.
5. Discuss problem framing. Compare and contrast it with range of negotiability for attributes.
6. Monika provides the following ordering of outcomes: $x_1, \ldots, x_5$. Is she consistent with the five rules?
   a. $x_1 \succ x_2 \equiv x_3 \succ x_4 \equiv x_5$
   b. $x_1 \succ x_2 \equiv x_3 \equiv x_4 \equiv x_5$
   c. $x_1 \succ x_2 \succ x_1 \succ x_4 \succ x_3 \succ x_4 \succ x_5$
   d. $x_1 \succ x_2$, $x_2 \succ x_4$, $x_4 \succ x_3$, and $x_3 \succ x_4$
   e. $x_1 \succ x_2$, $x_3 \succ x_4$, and $x_3 \succ x_4$
7. For case (a) in problem 5, Monika provides the following preference probabilities for the intermediate options. Is she consistent with the five rules?
   a. $P_{pref}(x_2) = 0.7$, $P_{pref}(x_3) = 0.7$ and $P_{pref}(x_4) = 0$
   b. $P_{pref}(x_2) = 0.7$, $P_{pref}(x_3) = 0.5$ and $P_{pref}(x_4) = 0.5$
   c. $P_{pref}(x_2) = 0.2$, $P_{pref}(x_3) = 0.2$ and $P_{pref}(x_4) = 0$
   d. $P_{pref}(x_2) = 1$, $P_{pref}(x_3) = 1$ and $P_{pref}(x_4) = 0$
   e. $P_{pref}(x_2) = 0.7$, $P_{pref}(x_3) = 0.7$ and $P_{pref}(x_4) = 0$
8. What is the difference between aleatory and epistemic uncertainty. Give examples of both.

9. For the following description of uncertainties, identify if they are aleatory or epistemic.
   a. Result of a coin flip
   b. Strength properties of a newly discovered fiber composite
   c. Outcome of an election
   d. Weather
   e. Wind velocity and direction exactly 2 months from now at your present location
   f. Length of a Boeing 747 aircraft for someone who does not know the true value
   g. Outcome of a chess game
10. What is the difference between a value function and a utility function? Give examples of both for some attributes of your choice.
11. How important is it for a decision-making method to incorporate preferences under uncertainty?
12. What is the certainty equivalent of a decision? Your friend Cody's utility function is given by $U(x) = x^2$ for an attribute x. What is his certainty equivalent of a decision situation that gives a normally distributed x with mean 2 and standard deviation 0.5?
13. What is certainty effect? How does it affect utility function assessment, and what is the way around it?
14. Danielle is indifferent between the following two lotteries. Assuming that she prefers more money to less, determine her utility function.

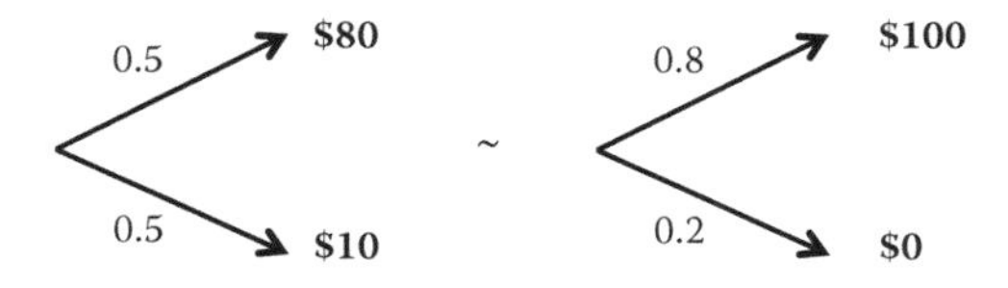

15. A nuclear plant needs to be set up in the outskirts of a big city. The plant needs to be close to the city if the I^2R and other logistical losses are to be kept to a minimum, which amount to $1 million per km distance per year. Finally, the plant has a meltdown probability in the next 30 years of 0.02, which, if it happens, can cause loss of life and property in dollar amounts of 3 billion (adjusted for inflation) that linearly decreases with distance from the city and drops to zero at 20 km. If the utility function over money (over 30 years) is exponential with risk tolerance equal to 50 million, determine where the plant should be located if the inflation adjusted profit per year is $20 million.

16. A manufacturer is considering integrating all components of a car bumper into an integrated snap-on design. The main implication he is considering is that it will incur a 20% reduction in assembly time (t) but a 10% increase in design and manufacture cost (c), from baseline values of 27 minutes and $75 per car. If he is indifferent between the baseline values and the values after the change, fit a value function to his preferences.
17. The manufacturer in problem 16 now tries to find his utility function over the value function he found earlier. Determine his utility function if he is indifferent between the sure option and the lottery below.

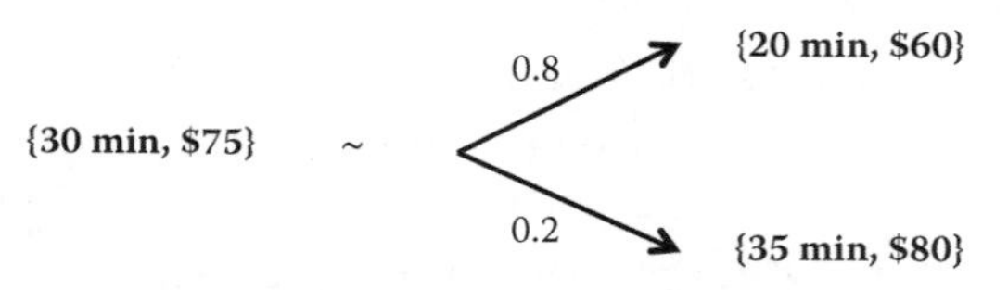

18. What are the three kinds of risk attitude a decision maker can have? What is the implication of each?
19. How will you determine a decision maker's risk attitude using the second derivative test? What can you do if his or her utility function is not differentiable?
20. The probability distribution of profit from sales of a yet to be designed product is normally distributed with mean 10 M and standard deviation 2 M. This can be turned into a deterministic profit of 12 M if the company secures a contract with another company, but procuring the contract will cost 2 M. Now consider the following utility functions and determine in each case if the company should choose to get the contract.
    a. $U(P) = P^{0.5}$
    b. $U(P) = 0.34P^{1.5} + 176000$
    c. $U(P) = \sin(0.000003P)$
21. A remanufacturer has 20 lawn mowers to assemble using used parts. Each lawn mower has 10 components and makes her a $1,040 profit. Parts availability is an issue. For each component the probability mass function is {0.1, 0.2, 0.3, 0.3, 0.1} that she will collect 18, 19, 20, 21, and 22 components, respectively. For each lawn mower she fails to assemble, she incurs a loss of $2,000 in lost opportunity and market image. If her utility function is exponential with risk tolerance equal to $20,000, find her expected utility. Should she sell the remanufacturing project to another company for $12,000?
22. A tire manufacturer has a stringent quality control program. Tires that do not make the most stringent requirements are binned into a second-best set and are then marketed as a different, lower-quality brand at

a lower price. To measure the quality of the tire, the manufacturer defines a metric t that lies between 0 and 1, 1 being the best. After analyzing the manufacture process and performing some tests, he realizes that the tires' metrics lie between 0.6 and 1 and follow a uniform distribution. The probability of failure of a tire during normal use is a function of the value of the metric and is given by $\frac{1-t}{20}$. A tire marketed as a top-quality tire can result in warranty costs of $2,000 per tire if it fails, and a tire marketed as a lower-quality tire results in a warranty costs of $200 per tire if it fails. The profit from the top-quality tire is $100 per tire, and that from the second-best set is $30 per tire. Where should he define the boundary of the two quality levels (in terms of t) if he plans to sell 100,000 tires and his utility function for money is given by $U(x) = 1 - e^{-x}$, where x is in millions of dollars.

23. During the product development process there is huge uncertainty in the future performance of the product as well as its potential market. Carefully delineate the steps you would take, using decision analysis, to make decisions regarding which design to pursue.
24. For Example 3.7, if two locations C and D have normally distributed wind speeds with means 15 and 20 m/s and standard deviations 2 and 4 m/s, which location should be chosen?
25. What is a Pareto front? Why is it used in design decision making?
26. What is an indifference curve? How will you draw indifference curves using utility functions? How will you get utility functions (if possible) using indifference curves?
27. What are the independence conditions associated with multiattribute problems?
28. What should one do if independence conditions are not satisfied?
29. A decision maker is indifferent between the following tuples of two attributes (both are to be maximized). Fit a value function.

Attribute Tuple 1	Attribute Tuple 2
{3, 12}	{3.5, 11}
{2.5, 10}	{2, 13}
{3, 8}	{3.1, 7.2}
{2, 7}	{3.5, 2}

30. For the value function you found in the previous problem, compare the two options A = {4, 4} and B = {5, 5}.
31. Normalize the value function found in problem 29, if the ranges of negotiability of the two attributes are [1, 8] and [1, 20], respectively.
32. When must we define utility functions over value functions?

4

Reliability Engineering

4.1 Performance as a Function of Time

Most systems, whether man-made or natural, degrade with time. While time itself is not liable for the degradation, it is the relevant phenomena that take place in time that lead these systems to degrade in functionality. The degradation can be caused by wear and tear from regular use, operating conditions that are harsher than anticipated, and last but not the least, design and manufacturing flaws. When the degradation reaches a certain critical level, *failure* occurs. A system fails when it can no longer perform its intended function satisfactorily. The end user usually defines the failure event(s); therefore, what counts as failure to one end user may not be so for another. It is possible to define failure in many different ways, ranging from not meeting cosmetic requirements to a major disaster that causes loss of property or life. Needless to say, a formal analysis of these events and their likelihood is needed so that failure can be avoided, or at least its likelihood of happening reduced.

Reliability engineering is the field of engineering that helps understand the system degradation phenomenon and provides a way to take corrective actions. Reliability concerns itself with learning the scenarios where a system fails, what the likelihoods of these scenarios are, the correlation between them, and finally, how much it will cost to either correct the failures or design the system so that the failure probability is minimized. Therefore, reliability engineering helps better plan the design, manufacture, and maintenance of engineering systems. It also helps prescribe operating conditions under which the system is most likely to perform its intended function. Higher reliability results in safe systems, lower repair and running costs for the customer, and also lower warranty costs for the manufacturer. Higher-reliability systems also have the intangible effect of creating a favorable image for the manufacturer.

Failures are rarely simple events. Many partial failures accumulate to cause system failure. Which partial failures are more likely to cause system failure and which are not must be understood if we want to understand system reliability. In such cases the *architecture* or *topology* of the system must be known.

Systems are divided into components that are functionally connected,* and these functional connections define the architecture or topology of the system. Needless to say, reliable components, when implemented in a good architecture, lead to reliable systems. A reliable system is, by definition, less likely to fail, a critical consideration in major installations such as dams and nuclear power stations.

Reliability texts define *reliability* as the probability that a system will perform its intended function for a specified amount of time under stated operating conditions (Kapur and Lamberson, 1977). Such probabilistic interpretation is very useful because operating conditions and system responses cannot be guaranteed with certainty. Such a definition is also mathematically rigorous. Given the operating conditions, component reliabilities, and system topology, one can find the probability that the system will perform its intended function. The manufacturer can also decide, based on this information, what warranty terms to offer and the expected warranty costs.

4.1.1 Reliability and Decision Based Design

Reliability engineering is well within the scope of decision based design (DBD). Making a system reliable is a decision, so is investing money and effort into achieving reliability. There are some differences, though, which we will discuss in a little more detail in Section 4.5. Concisely, decision based design is focused on attributes of a system, while reliability is concerned with the binary events of success and failure for not meeting the targets in these attributes. Consequently, if a formal design process utilizing DBD principles can evaluate all outcomes and probabilities and propose the best solution, there is no real need for the surrogate metric of reliability. There are, however, situations where it makes sense to focus merely on success/failure events and make reliability the objective of the design process. We will return to this discussion later in the chapter.

4.2 Reliability Engineering Concepts

In this section we familiarize ourselves with common concepts used in reliability engineering. Standard reliability engineering notation is used. Wherever needed, real-life examples are provided to help the reader understand the concepts better.

* Systems are sometimes composed of products, which are themselves composed of components and subcomponents. For simplicity, in this chapter we will call the units composing a system, components.

4.2.1 Reliability Function

A reliability function is a mathematical expression that takes into account an engineering system's construction and susceptibilities and provides the probability that it will survive beyond a prespecified time under stated conditions. The system's construction and susceptibilities under stated conditions become embedded in the constant *parameters* in the reliability function. Therefore, once the parameters have been determined, the reliability function takes the time as input, and outputs the probability that it will still be working at that time, that is, the probability that it will fail only *after* that time. As expected, this function decreases monotonically with time for most products. Products such as antiques and wines that increase in value with time are, of course, exceptions and are not discussed in this book.

Consider a frequentist's* view of system failure. Imagine that you have a collection of systems. They cannot all be expected to fail at the same time. Some will fail quite early, some will fail at or near the expected lifetime of the product, while some will fail much later. Therefore, the time to failure of the system will follow a distribution. And the probability that a given system will last longer than time t is the number of surviving components at time t divided by the total number of systems when we started. Mathematically,

$$R(t) = \frac{N_{surviving}(t)}{N_{total}} = \frac{N_{surviving}(t)}{N_{surviving}(t) + N_{failed}(t)} \tag{4.1}$$

Translated into probability notation, this can be rewritten as

$$R(t) = \mathrm{P}(T > t) \tag{4.2}$$

where T is the time to failure of a component. We can rewrite the equation as

$$R(t) = 1 - \mathrm{P}(T \leq t) \tag{4.3}$$

The second term of the right-hand side of the equation is essentially the cumulative density function of the variable T; therefore,

$$R(t) = 1 - F_T(t) \tag{4.4}$$

* The term *frequentist* refers to individuals who believe who probability is simply the relative frequency of an outcome, also sometimes referred to as objective probability. Bayesianists believe there is no such thing as objective probability and that probability is the degree of belief in the decision maker's mind that an outcome will occur.

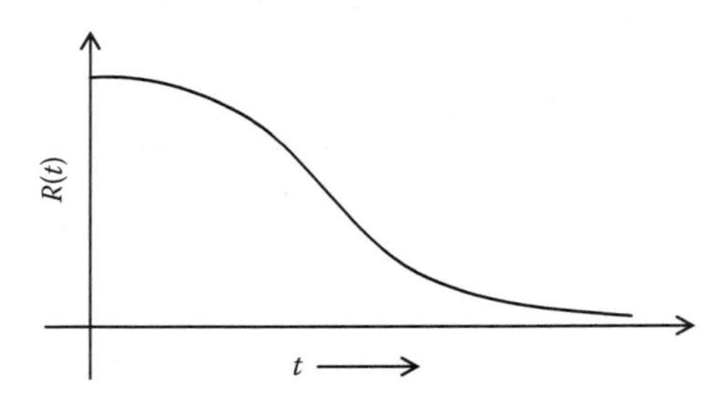

FIGURE 4.1
A typical reliability curve.

In other words, reliability is 1 minus the CDF of the time to failure of the system. Figure 4.1 shows a typical reliability function of a system.

Many times only the density function is known from failure events; the above equation can be rewritten as

$$R(t) = 1 - \int_0^t f_T(x)dx \tag{4.5}$$

where $f_T(x)$ is the probability density function (pdf) of the time to failure, T. A closely related concept to reliability is mean time to failure (MTTF)* of a system. MTTF is the expected time at which the system fails. This is true, however, only in the statistical sense; that is, it is not necessary that the system (or even most systems) fail around the MTTF. We can calculate the MTTF value using the pdf of the time to failure as

$$MTTF = \int_0^\infty tf_T(t)dt \tag{4.6}$$

If various times to failure of systems of the same type are available, MTTF is simply their arithmetic mean. MTTF does not capture all the characteristics of a reliability function. For example, if half of a collection of systems fail at 10 hours and the rest at 10,000 hours, the MTTF is 5,005 hours. Not only do none of the systems fail at 5,005 hours, but the number is also not very representative of the time to failure. Despite this shortcoming, it is a widely used metric for reliability and commonly used in product design specifications. One must realize therefore that one might be missing critical information about the time to failure distribution when working with only MTTF.

* Sometimes abbreviated as MTBF, which stands for mean time *between* failures.

The mean of the performance functions, $g(\mathbf{x})$, is simply its value calculated at the mean value of the design variables; thus,

$$\mu_g = g(\mu_{x1}, \ldots, \mu_{x2}) \tag{4.14}$$

And its variance is given by

$$\sigma_g^2 = \sum_i \sum_j \frac{\partial g}{\partial x_i} \frac{\partial g}{\partial x_j} Cov(x_i, x_j) \tag{4.15}$$

The above expressions are accurate only if design variables follow a normal distribution and $g(\mathbf{x})$ is actually linear in the design variables. The safety or reliability index, β, is defined using the mean and standard deviation of $g(\mathbf{x})$ as

$$\beta = \frac{\mu_g}{\sigma_g} \tag{4.16}$$

Therefore, the safety index is the number of standard deviations the failure boundary is away from the design mean. The performance functions also have an optimization interpretation; they act as constraints in the optimization problem. The essential complication is that the constraints are stochastic, and for each design we can find their probabilities being violated. We then select the one that maximizes the objectives while minimizing the probability of these violations. Such design problems are studied under the context of reliability-based design optimization, usually abbreviated RBDO. Interested readers are referred to reliability engineering texts for a detailed treatment of these topics.

The treatment in this section can predict the failure probability for a given design and given limit state functions. Recall that the reliability function also includes a time element. To find the time-dependent reliability of a system, one would require knowing how the limit state functions as well as the statistics of the design variables' change with time. If time degradation phenomena can be captured correctly, time-dependent reliability can be found.

4.2.4 Weibull Distribution

The distribution that closely models the time to failure for most systems is the Weibull distribution. The pdf and CDF of the Weibull distribution are given by

$$f_T(t) = \frac{b}{\lambda}\left(\frac{t}{\lambda}\right)^{b-1} \exp\left[-\left(\frac{t}{\lambda}\right)^b\right] \tag{4.17}$$

$$F_T(t) = 1 - \exp\left[-\left(\frac{t}{\lambda}\right)^b\right] \tag{4.18}$$

The parameter λ is called scale parameter and b is the shape parameter. Notice that these are the distribution functions of the time to failure. The reliability function of a system whose time to failure follows the Weibull distribution is given by

$$R(t) = 1 - F_T(t) = \exp\left[-\left(\frac{t}{\lambda}\right)^b\right] \tag{4.19}$$

The hazard rate is

$$h(t) = \frac{b}{\lambda}\left(\frac{t}{\lambda}\right)^{b-1} \tag{4.20}$$

The mean time to failure of a Weibull system is given by (Γ is the gamma function)

$$\text{MTTF} = \lambda\Gamma\left(1 + \frac{1}{k}\right) \tag{4.21}$$

Another distribution that is sometimes used in place of Weibull is the normal distribution.

Example 4.1

The scale and shape parameters of the Weibull distribution followed by the time to failure of an electronic transistor are 30,000 hours and 2. What are its reliability and hazard rate at 10,000 hours?

SOLUTION

Reliability at 10,000 hours is given by

$$R(10000) = \exp\left[-\left(\frac{10000}{30000}\right)^2\right] = 0.895$$

Hazard rate is given by

$$h(10000) = \frac{3}{30000}\left(\frac{10000}{30000}\right)^{2-1} = 3.33 \times 10^{-5}$$

4.3 Failure Modes and System Reliability

System reliability is a function of the reliabilities of the components composing it. To determine it, though, we will need to know, in addition to component reliabilities, which combination of component failures leads to system failure and which does not. Each combination of failed components that also leads to system failure is called a *failure mode* of the system. To determine the failure modes, one needs to understand connectivities between components. Let us start with a two-component system. If failure of either of the components leads to system failure, the components are said to be connected in series, while if failure of both the components is required to make the system fail, then they are said to be connected in parallel. Figure 4.4 shows the block diagram representations (which we will discuss shortly) of two components connected in (a) series and (b) parallel. The idea can be easily generalized to many components in series or in parallel.

The failure modes of the series system in Figure 4.4 are {1}, {2}, and {1,2}. However, the only failure mode for the parallel system is {1,2}. Exhaustive collection of failure modes constitutes what is known as the topology of the system. There are many ways to represent the topology of the system, visually or otherwise. In this book we present two representations: block diagram representation and fault tree representation.

4.3.1 Block Diagram Representation

Block diagram representation of a system topology resembles an electric circuit, as shown in Figures 4.4 and 4.5. In this representation, two components are in series if they are connected end to end. The reliability of a series

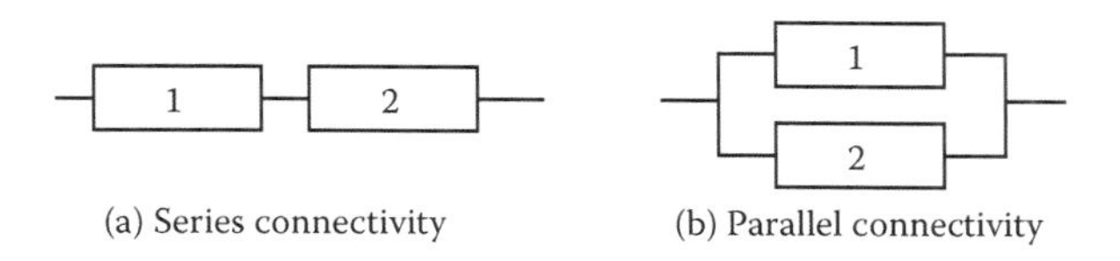

FIGURE 4.4
Two basic types of connections between pairs of components: (a) series and (b) parallel.

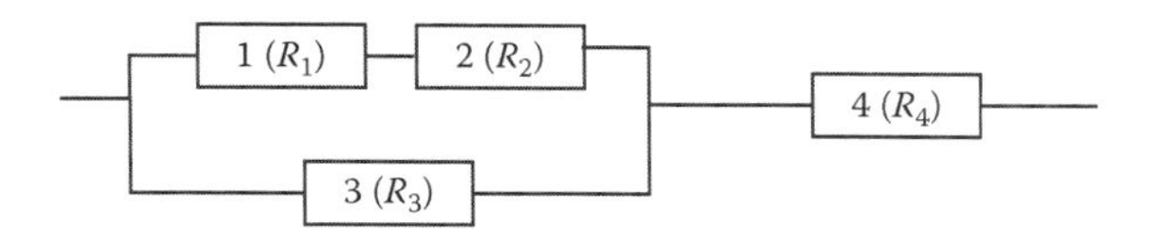

FIGURE 4.5
Reliability block diagram of a simple system.

system is the product of the individual component reliabilities, since all components need to survive for the system to function. Similarly, reliability of a parallel system is the probability that at least one component survives, i.e., 1 minus the product of 1 minus their reliabilities. Most systems, when broken down to individual components, show a combination of series and parallel character. For example, consider the system shown in Figure 4.5. A block diagram lends itself very well to visual inspection of failure. For example, if there is a path in the diagram from left to right (or from right to left) then the system is working.

To evaluate the reliability of the system, we work with pairs of adjacent components and identify the series or parallel connections between them. For example, in Figure 4.5, components 1 and 2 are in series, and therefore the reliability of that subsystem is R_1R_2. This subsystem is in parallel with component 3, and therefore the reliability of the whole subsystem comprising components 1, 2, and 3 is equal to

$$1-(1-R_1R_2)(1-R_3) \tag{4.22}$$

Finally, realizing that component 4 is in series with this subsystem, we get the overall system reliability as

$$R_{system} = R_4 \cdot (1-(1-R_1R_2)(1-R_3)) \tag{4.23}$$

Dividing the system into subsystems where each subsystem can be identified as a series or parallel system allows us to find the reliability of most commonly encountered engineering systems. The system shown in Figure 4.5 can fail in many ways depending on which components fail. These failure modes can be exhaustively acquired from the block diagram of the system. A truncated list involves the component combinations $\{1,3\}$, $\{2,3\}$,$\{1,2,3\}$ and $\{4\}$.

It must be kept in mind that some systems cannot be reduced into sequences of series and parallel subsystems. Figure 4.6 shows an example of such a system. The general approach in such cases is to define mutually exclusive failure events and find their probabilities. Kapur and Lamberson (1977) provide details on this approach. One can also approximate the reliability of such systems using simulation.

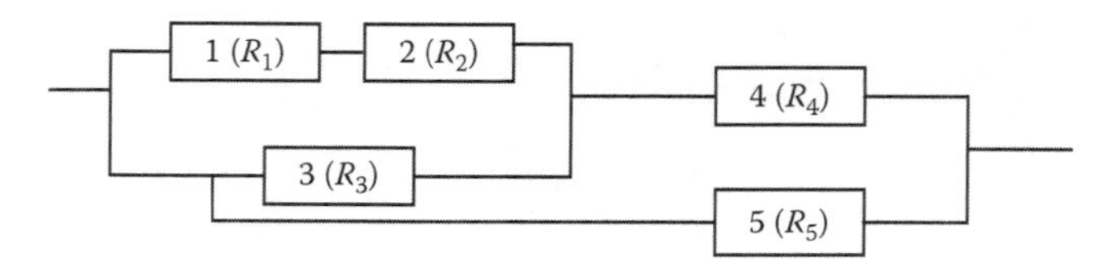

FIGURE 4.6
Reliability block diagram of a non-series–parallel system.

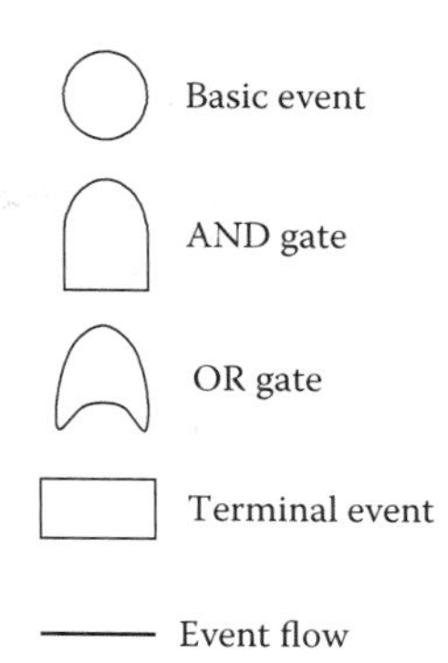

FIGURE 4.7
Basic elements of a fault tree.

4.3.2 Fault Tree Analysis

System topology can also be represented using a fault tree. A fault tree propagates the failure events of components using Boolean operations of AND and OR. The failure events of working/nonworking components can be represented using the binary state values of 1 and 0. An AND gate acts like the intersection operator in that all the inputs should be 1 for it to report a 1, while an OR gate reports 1 if at least one of the inputs is 1. Figure 4.7 shows the basic elements of a fault tree.

Figure 4.8 shows the fault tree corresponding to the system in Figure 4.5. There are four basic events corresponding to the four component failures.

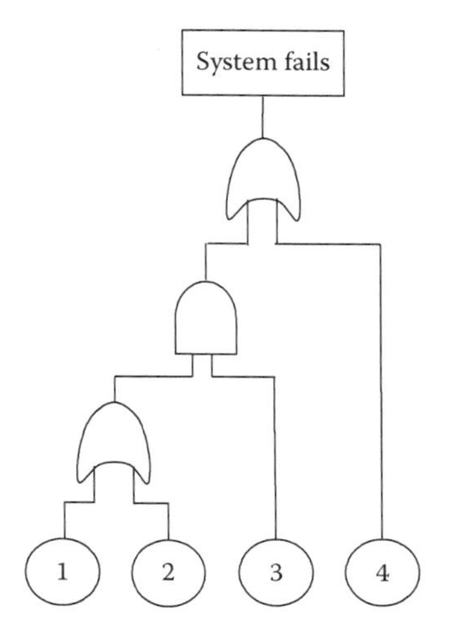

FIGURE 4.8
Fault tree for the system shown in Figure 4.5.

Components 1 and 2 feed into an OR gate signifying that the failure event will be transmitted if even one of the components fails. The output of the OR gate feeds into an AND gate together with component 3. This is equivalent to the failure information in Figure 4.5 where component 3 provides redundancy to the series combination of components 1 and 2. Finally, the output from the AND gate feeds into another OR gate along with component 4 because of the series combination, as shown in Figure 4.5.

Notice that both block diagrams and fault trees represent the same information. Block diagrams have the advantage that they can be easily implemented in a system simulation tool. It is easier to write a computer code that evaluates the Boolean logic than the inferences required in a block diagram.

4.4 Reliability of Repairable Systems

Most of the discussion in this chapter has focused on unrepairable systems, or systems that are discarded when they fail. Most real-life systems are not like that; they are repaired (fully or minimally) and put back to use. This, however, complicates the definition of reliability, particularly when the maintenance and repair strategy is allowed to vary. Thinking strictly in terms of the reliability function, every upgrade interrupts the natural decline of the reliability function and increases it to the value of 1, as shown in Figure 4.9. This is because regardless of full or partial (e.g., minimal) repair, a system is only put back to use after it has been brought back to working condition. The slope of the reliability curve, however, depends on the type of repair, because in the repaired system there are still some old working components or some nonworking components (in the case of partial repair). The repair makes the overall reliability curve nonmonotonic and the classical meaning

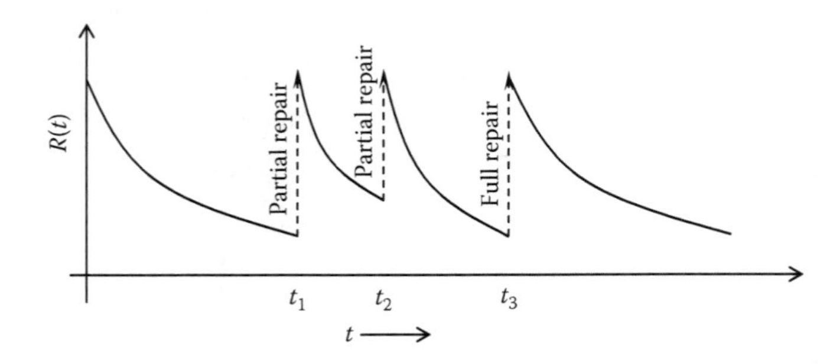

FIGURE 4.9
Representation of the reliability of a repairable system as a function of time. Many unavoidable inconsistencies are encountered if we use a classical definition of reliability.

of reliability (probability that the system has not failed before a given time) is lost. Within a segment of the curve, the concept of reliability makes sense only if we reset the definition at each repair time. Combine this with the fact that the reliability value at any given time t can be arbitrarily set, depending on when repair is performed, and we have further issues. Clearly to measure performance of a repairable system we need different metrics than the simplistic classical notion.

We discussed mean time to failure (MTTF) earlier in the chapter. For a system subjected to many sequences of operation-repair cycles, MTTF provides a useful metric for performance of the system. If we also include mean time to repair (MTTR), we can define availability as

$$Availability = \frac{MTTF}{MTTF + MTTR} \quad (4.24)$$

which is the fraction of time a system is operational (notice that MTTR is assumed to cover all downtime). Notice that both availability and MTTF suffer from the same issue. They fail to report the full distribution of the operation and repair times and can lead to either overly conservative or overly aggressive estimates of system performance. For example, a system might fail after every hour of operation, but if the repair time is 6 minutes, the system has greater than 90% availability. It is hard to get any meaningful service out of such a system. Similarly, a system can have a high variance in its operation times, but MTTF will not reflect that. A better way to look at MTTF is to define a minimum failure-free period (MFFP). MFFP should be defined with a given probability. Therefore, for a probability p, MFFP is $(1-p)\times$ 100th percentile of the operation time distribution. Mathematically, if $F_T(t)$ is the CDF of the operation times, MFFP with probability p is given by

$$MFFP = F_T^{-1}(1-p) \quad (4.25)$$

MFFP and MTTF can together provide a very good picture of system reliability.

4.4.1 Effective Age of Repaired Systems

An issue recognized with repaired systems is that they are not perceived to be of as much value as new systems. In this section we will discuss the method of effective age developed by Pandey and Thurston (2009) to report performance of such systems. The underlying concept is that when reporting the performance of such systems, we can use the units of time, describing how old a never disassembled (unrepaired, working) system has to be to have the same reliability as the given repaired system. There are nine desiderata the metric of effective age $(\bar{t})$ should satisfy:

1. The method should be able to report the overall performance level (effective age) of the system even if the components have different ages (allowing partial repair).
2. The method should be able to include subjective information about the components' value degradation acquired from end users or experts.*
3. If all the components are the same age, the method should reflect that fact in the effective age of the system (it should be the same as the age of the components).
4. The measure should account for the relative importance or criticality of the components.
5. A high-performing component should compensate for the low performance of others, especially if it is critical as determined by the system topology.
6. Each component should have a cutoff age after which no compensation from other components is sufficient to make the system acceptable.
7. A system's effective age should not be less than that of its newest component or greater than its oldest component. This is a general case of desideratum 3.
8. The method should include information about system failure modes and be able to accommodate changes in them.
9. The effective age of the system should increase if the ages of individual (nonredundant) components increase.

Let us consider a system with n components, each with a reliability function $R_i(t_i)$, while the functional $F(.)$ combines the reliabilities of components to give the reliability of the system (i.e., the failure modes are embedded in F). The reliability of the system is given by

$$R(t_i,...t_n) = F(R_i(t_i),...R_n(t_n)) \tag{4.26}$$

Notice that the components are allowed to be different ages, as one would expect in a repaired system. We can also rewrite Equation 4.26 for the special case when all components are the same age. Let $R^S(t)$ be the function that $R(t_i,...,t_n)$ reduces to when all the components are the same age (superscript S signifies same age).

$$R^S(t) = F(R_i(t),...R_n(t)) \tag{4.27}$$

* While the method is capable of incorporating subjective information, we will not be discussing it here. Interested readers are referred to the original paper by Pandey and Thurston (2009).

From desideratum 3 we know that the effective age of the system should be the same as that of its components if they all are the same age. We have

$$t = R^{s^{-1}}(F(R_i(t),\ldots,R_n(t))) \tag{4.28}$$

The effective age of a system with components of different ages, $\bar{t}$, would be given by Equation 4.27:

$$\bar{t} = R^{s^{-1}}(F(R_i(t_i),\ldots,R_n(t_n))) \tag{4.29}$$

In other words, the method compares a system's reliability with that of a system that has never been disassembled and reports the corresponding effective age. When an end user acquires a repaired system, he or she makes assessments of whether it would provide acceptable service in the future, based on how old the individual components are. The effective age metric quantifies this assessment and provides an anchor, a never disassembled system, with which to compare the system. The method therefore takes one away from a mathematical definition of reliability and uses an easily relatable unit of time.

Depending on the functional form of R's and F, it can be difficult to find a closed-form expression for the inverse of R^S. The workaround is that one can numerically calculate the value of R^S for different values of t and store it in a data file. When finding the effective age, the value of $\bar{t}$ is read back from the file corresponding to the reliability of the system calculated using Equation 4.29. Since the same reliability of the system can be achieved by different combinations of component ages, different systems can have the same effective age (as expected).

First-order approximations can also be used to simplify the calculation of the effective age, $\bar{t}$, in certain cases. Let us consider the partial derivatives of $\bar{t}$ with respect to individual t_i's:

$$d\bar{t} = \frac{\partial \bar{t}}{\partial t_1}dt_1 + \frac{\partial \bar{t}}{\partial t_2}dt_2 + \cdots + \frac{\partial \bar{t}}{\partial t_n}dt_n \tag{4.30}$$

The individual partial derivatives can be called the criticalities of the corresponding components. These partial derivatives can be used to directly approximate $\bar{t}$, if its value is known for nearby values of t_i's. This can make the calculation computationally simple. It can also be easily shown that the criticalities add up to 1 when all components are the same age. The calculations involved in the linear approximation are much simpler and can be used to quickly estimate effective age if minor upgrades are made to the system. The criticality information can also be helpful in determining on which component one should expend energy and efforts to improve system performance.

TABLE 4.2

Criticalities for PC Components

Component	Criticality
Monitor	0.065
Keyboard	0.03
Hard drive	0.25
CD-ROM	0.045
Processor	0.325
Power supply	0.035
RAM	0.15
Video card	0.1
Sum	1

Example 4.2

Component criticalities (partial derivatives as in Equation 4.30) of the major components of a personal computer are given by Table 4.2.

Now imagine that you have a 4-year-old computer and want to upgrade one component, either the processor or the video card. How old do they have to be if you want to be able to sell the upgraded computer to someone who wants the effective age to be at most 3 years.

SOLUTION

We solve the following equation to find the age of the component that will bring about the required change in the effective age of 1 year (4 to 3):

$$4-3=\frac{\partial \bar{t}}{\partial t_i}(4-t_i)$$

For the processor, we have

$$1=0.325(4-t_{processor})$$

$$t_{processor}=0.92 \text{ years}$$

Therefore, if the existing 4-year-old processor is upgraded (replaced) with a processor that is 0.92 year old, the effective age of the computer will come down to 3 years. Now for the video card we have

$$1=0.1(4-t_{videocard})$$

$$t_{videocard}=-6 \text{ years}$$

which implies that it is not possible to achieve the required change in the effective age of the computer by replacing the video card, even if we do so with a brand new card. In fact, if the linear approximation holds, it will have to be from 6 years in the future!

4.5 Reliability–Cost Trade-Offs

In the introduction section to this chapter, we discussed how reliability comes at a cost. Regardless of how extreme the outcomes of failure of a system are, no system is (or can be) designed with perfect reliability. In fact, the closer one gets to a reliability of 1, or a similar high value over the lifetime of a product, the cost of designing and manufacturing such a system increases exponentially. For noncritical applications the cost of the overall system gives a good idea of where we should stop reliability efforts. If it is going to be easier to replace the system than to design it with higher reliability, there is no need to increase reliability. Redundancy with standby systems generally achieves higher reliability than making one system more efficient.

Even in critical applications such as nuclear power plants or airplanes, reliability–cost trade-offs are routinely made. Failure of these systems can result in substantial loss of life and property. Allowing a small probability of failure to reduce cost seems counterintuitive, but it is not necessarily so. We take many risks in our daily lives that equal or exceed the probability of suffering catastrophic failure of nuclear power plants or airplanes. It is a well-known fact, for example, that having a fatal accident while driving is significantly higher than dying in a plane crash over the same distance traveled. Extreme precautions are taken in the design of airplanes and nuclear power plants. The latter are usually built in remote areas with many backup systems built in to avoid failure and overheating of spent fuel. In fact, decision analysis has been used successfully in making nuclear plant siting decisions. Keeney and Nair (1975), for example, show how multiattribute decision analysis can be used to choose between candidate sites.

4.6 Notes on Decision Based Design

Reliability engineering concerns itself with the binary events of failure and success. More succinctly, it models the probability that failure events will not occur. Decision based design goes beyond that. Decision based design will look at what happens when these failures do occur. Clearly, the impact of the failures on the system attributes of concern can range from benign to those with extreme repercussions and include anywhere in between. If probabilities of different failure events and the attribute levels are available, one could simply maximize the expected utility from these attributes. After all, in a decision situation, we enumerate all possibilities (including bad ones) and their relevant probabilities. Consequently, maximizing the expected utility from the design decision should do the trick. Decision based design therefore gives a richer paradigm by using an integrated framework and can in theory replace reliability engineering. Consider an airplane design problem. Catastrophic frame failure is definitely more critical than an engine surge event, which is more critical than decreased fuel efficiency, which in turn is more critical than uncomfortable seats. Obviously, if each of these failures were given equal weight, we would end up with a suboptimal design. Even if different weights are assigned to these failure probabilities, we are still maximizing probabilities. Maximizing a probability (reliability is a probability) has also been considered to be intuitively incorrect, because in the Bayesian interpretation, probability is a degree of belief. How can we optimize a degree of belief?

Despite these issues, reliability engineering still has significant practical value. In complex systems involving thousands of components, it is not possible to enumerate all scenarios. The system must be broken down into individual components and subcomponents. In the decomposed form of the system, it makes sense to talk about likelihood of meeting a target, particularly when there are many standardized parts (e.g., individual screws) whose failure probabilities of not meeting the targets are known. When a manufacturer outsources some components to another company, the specifications cannot be in the form of "maximize power output" and "minimize weight." The outsourced component should be assemblable with the rest of the product and have precise performance characteristics. In such cases, prescribing a reliability function or even a mean time to failure is practically more useful than a DBD approach.

Problems and Exercises

1. What is reliability engineering? Why should one study reliability engineering?
2. How is the reliability function of a system defined? How is reliability related to the CDF of the time to failure?
3. What is mean time to failure (MTTF)? Why does it not capture all the characteristics of a system's reliability?
4. What is the most commonly used distribution for characterizing the reliability of a system? Write expressions for failure rate and mean time to failure.
5. The scale and shape parameters of the Weibull distribution followed by the time to failure of a generator are 2 days and 4. What is its reliability and hazard rate after 60 hours of continuous operation? What is its MTTF?
6. The reliability of a component is 0.94 at 10 hours and 0.8 at 15 hours. Find the shape and scale parameters for the Weibull distribution that will lead to these values.

7. The times to failure of 50 computer mice in thousands of hours are given in the following table. What is the MTTF? What is the reliability at 3,000 hours?

0.012	5.358	0.288	1.677	2.126
6.638	6.167	6.996	1.437	4.626
1.705	0.751	0.150	6.675	6.397
3.621	0.146	5.142	3.000	3.301
0.698	4.661	3.244	0.190	2.100
0.247	0.133	3.473	3.977	3.423
4.838	4.500	1.342	0.105	0.022
1.019	0.207	2.179	0.123	0.189
6.103	1.673	0.076	3.081	3.304
0.001	3.977	1.162	0.573	6.434

8. What is a performance (limit state) function?
9. Differentiate between FOSM and SOSM methods.
10. The reliability at 1,000 hours of eight components is given in the following table. Find the reliability of the systems shown.

Component Number	Reliability	Component Number	Reliability
1	0.60	5	0.66
2	0.90	6	0.85
3	0.99	7	0.25
4	0.88	8	0.95

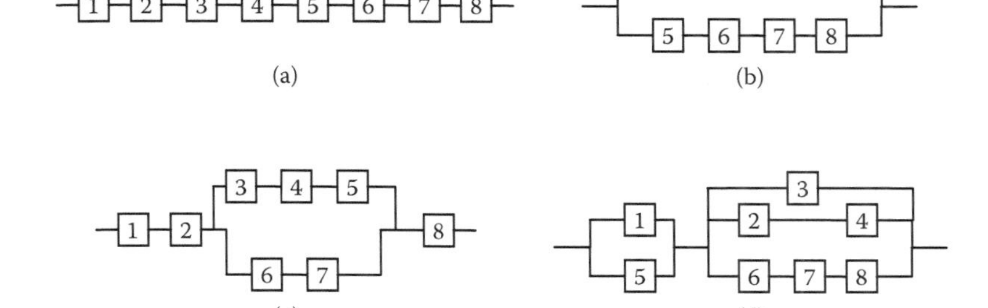

(a) (b) (c) (d)

11. What is a fault tree? Compare and contrast a fault tree with a reliability block diagram.
12. Why is it hard to define reliability of repairable systems in the traditional sense?

13. Discuss a few common metrics used to measure the reliability of repairable systems.
14. The failure and repair times of a repairable system for the first 15 runs are given below (in days). What is the availability of the system? What is its minimum failure free operating period (MFFP) with 80% probability?

Running Time	Repair Time	Running Time	Repair Time	Running Time	Repair Time
4.06	0.29	1.19	0.02	2.88	0.47
3.76	0.33	4.89	0.01	3.37	0.97
2.78	0.64	1.88	0.21	0.36	0.58
1.51	0.76	4.15	0.02	2.63	0.98
4.13	0.91	4.20	0.74	0.49	0.67

15. How does reliability engineering relate to decision based design?

5

Design Optimization

5.1 Modeling an Engineering System

One of the biggest contributions of computers to the field of engineering is the ability to simulate engineering systems on them. This has allowed us to solve simulated versions of problems involving the systems that were traditionally considered intractable. On a computer, a problem can be solved many times over, with different design parameters and under different simulated operating conditions. Doing so allows us to understand the system's behavior without resorting to expensive tests. Also, if needed, one can find the design parameters that maximize an objective (or performance) function. A wide variety of engineering design problems understandably involve extremization (maximization or minimization) of objective functions or attributes. Some examples include minimizing stress in a machine element, minimizing noise produced by an aircraft engine, and maximizing torque produced by a motor. The field of study that deals with these types of problems is called *design optimization*. Design optimization relies on both the mathematical model of the engineering system and the techniques that enable optimization of an objective, given such a model.

Unfortunately, computers cannot solve design optimization problems themselves; they need to be programmed with an optimization algorithm. Computers can then implement the algorithm and execute the steps many times over, if needed, or until certain criteria are met. Efficiency and accuracy of algorithms obviously have immense value, and this chapter is devoted to the understanding of basic concepts that make an optimization algorithm work. Optimization is an active field of study because, despite the immense value it provides, there are also many roadblocks to successful implementation. There is no cure-all method for optimization; that is, no single optimization algorithm can solve all engineering problems efficiently. The first and arguably most challenging roadblock to optimization is the need of a model of the engineering system. In the engineering design context, a model is defined as a mathematical description of the working of a system. An engineering model provides the values of output parameters of a system as a function of the inputs. Such models are hard to construct. At the one extreme, they could be simple quadratic functions that can be manipulated by hand. At the other extreme lie problems such as a million degrees of freedom finite element model that predicts whether a certain impact on an automobile will damage the passenger cage, or a comprehensive dynamic and thermal model of the environment that a connecting rod is subjected to in an engine, which can help us tell the effect of design changes without having to perform extensive experiments. While the system response to inputs in these difficult problems can be acquired using computers running complex finite element or computational fluid dynamics (CFD)* codes, the computational effort is large, and usually no closed-form expression exists. Lower-order approximations can be made, but their accuracy cannot be guaranteed. Generally, the better a model approximates the engineering system, the more complex it is, and hence the unavoidable trade-off in model accuracy and complexity.

Another roadblock to optimization is the inherent mathematical characteristics of an engineering model. The model exists as a representation of the system, not to enable or facilitate optimization. Properties such as differentiability or convexity of functions and constraints, while extremely desirable from an algorithmic perspective, are rarely satisfied. Many times it makes sense to sacrifice fidelity in an engineering model to enable efficient optimization. The solutions may not be accurate but they provide a good starting point subject to further refinement. Lower-order approximations (replacing the true model with a simple function such as a quadratic) have been used successfully in solving many engineering problems. Model validation and verification research actively deals with these types of problems. Other roadblocks to optimization include black-box type systems where no closed-form input–output relationship exists, many constraints—some of which may be unknown, the presence of many local optima, and so on.

For the purposes of this chapter, we will assume that we have a way of knowing the system response given the inputs. Some methods we will consider will require this model to be in the form of an explicit mathematical formulation. In the later parts of the chapter we will go over heuristics that can treat the system as a black box and do not need an explicit mathematical form (e.g., genetic algorithms and ant colony optimizers). We will also consider methods to generate a Pareto front over multiple attributes if it is not possible to combine them into one objective function.

5.1.1 How to Decide What to Optimize?

The objective of optimization sometimes occurs naturally in design problems, while at other times it is imposed externally. Minimizing deflection of a beam in structural optimization may be considered a natural objective. On the other hand, when designing a platform of products, using a component

* CFD is a field of engineering that studies fluid flow.

commonality index as an objective may not be immediately obvious. Also, very rarely do engineering design problems involve only one objective. Multiple conflicting objectives often exist, with varying degrees of importance. They often need to be combined into a performance or objective function. The standard form of an optimization problem is as follows:

$$\begin{aligned} \text{Minimize} \quad & f(\mathbf{x}) \\ \text{subject to} \quad & \mathbf{x} \in X \end{aligned} \tag{5.1}$$

In the above formulation $f(\mathbf{x})$ is the objective function that needs to be minimized by choosing the best values of variables within the vector $\mathbf{x} = \{x_1, \ldots, x_n\}^T$. The vector $\mathbf{x}$ is constrained to be contained in a set X called the *feasible set*. The feasible set is often written as the regions of design space that satisfy certain explicit criteria, known as constraints. The value of $\mathbf{x}$ within the feasible set that minimizes $f(\mathbf{x})$ is the optimal solution and is written as $\mathbf{x}^*$. Notice that there may be other values of $\mathbf{x}$ that find a lower value of f than $f(\mathbf{x}^*)$, but they may not be admissible because they are not contained in X. Conventionally, all optimization problems are written as minimization problems, but this does not result in any loss of generality because a suitable transformation (e.g., putting a negative sign in front of the objective) easily transforms a maximization problem into a minimization problem.

The holy grail of decision based design is complete integration of decision analysis and design optimization. To that effect, maximizing the multiattribute utility function of the decision maker therefore should be the objective of optimization. Decision analysis helps us identify utility functions, and optimization algorithms can help their maximization. As we saw in Chapter 3, identifying utility functions is easier said than done. Utility function assessments require carefully constructed questions that are posed to the decision maker. A mathematical function is then fit to the responses provided by the decision maker. In the context of optimization, it is a good idea to fit a smooth curve to the implied utility function, which is usually assessed only at discrete points. This is because a smooth differentiable curve allows for utilization of methods in classical optimization through the use of gradients and easily calculable Hessians, as we shall see later. In many real-life situations, surrogates such as cost, deflection, temperature, and modularity index are used because a utility function is not available. Such surrogates are justified when the effort required to assess the utility function will outweigh the benefits.

5.1.2 Constraints

Constraints restrict what designs can be realized. They may be obvious limitations on design; for example, mating parts in an assembly should

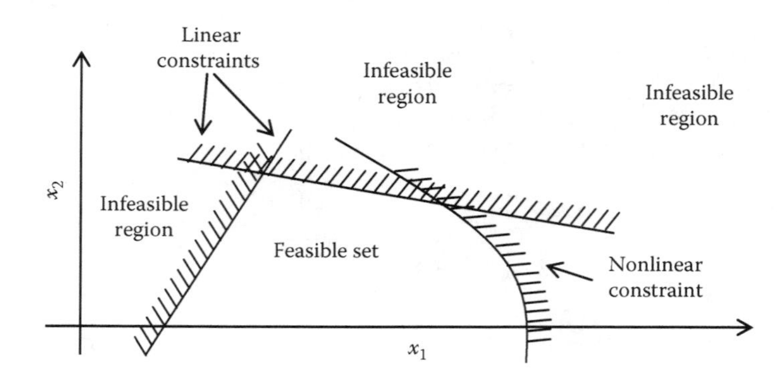

FIGURE 5.1
A graphical representation of two linear constraints and one nonlinear constraint in two dimensions.

have the same interface dimensions. Alternatively, they may have to do with the working of the system. Imagine a distributed design of a large commercial aircraft. To generate enough lift for takeoff, it is essential that the engines produce enough thrust within the takeoff run. This puts a lower limit on engine thrust. On the other hand, very strong engines are costly to develop, use more fuel, are noisy, and are generally larger in size. All of these provide an upper limit on the thrust generated. Any engine design, to be feasible, needs to satisfy these limitations, among many others. In the optimization context, these limitations are called *constraints* (Figure 5.1). Engineering design problems are usually severely constrained, which makes it difficult to maximize or minimize objectives because seemingly best solutions are very often found to be infeasible. Oftentimes, constraints also limit what algorithms can be used to solve the problems.

Constraints are generally classified into two types: equality constraints and inequality constraints. Equality constraints force some functions of the design variables to equal a constant. Inequality constraints require that certain functions of the design variables take a value less than a constant. Conventionally, equality constraints are represented with $h_i(\mathbf{x})$, where i indexes the different equality constraints. Similarly, inequality constraints are represented with $g_i(\mathbf{x})$, the subscript i again indexing the different inequality constraints. Equality constraints in general are harder to satisfy than inequality constraints; they also restrict the design space more. See, for example, Figure 5.1; if one of the linear constraints was actually an equality constraint, the solution will have to lie *on* the line.

To write constraints in a standard form, we modify them into the form of $\mathbf{h}(\mathbf{x}) = \mathbf{0}$ and $\mathbf{g}(\mathbf{x}) \leq \mathbf{0}$, where $\mathbf{h}(\mathbf{x})$ and $\mathbf{g}(\mathbf{x})$ are vectors of the two types of constraints. This modification can be done without losing generality, as can be seen with the following examples.

Example 5.1

Write the following constraints in the standard form:

a. $q(\mathbf{x}) = p(\mathbf{x}) + c$
b. $q(\mathbf{x}) = d$
c. $p(\mathbf{x}) \leq r(\mathbf{x}) - d$
d. $f(\mathbf{x}) \geq 0$
e. $r(\mathbf{x}) > d$
f. $r(\mathbf{x}) < z(\mathbf{x}) + d$

SOLUTION

a. Define $h(\mathbf{x}) = q(\mathbf{x}) - p(\mathbf{x}) - c$, the constraint becomes $h(\mathbf{x}) = 0$.
b. Define $h(\mathbf{x}) = q(\mathbf{x}) - d$, the constraint becomes $h(\mathbf{x}) = 0$.
c. Define $g(\mathbf{x}) = p(\mathbf{x}) - r(\mathbf{x}) + d$, the constraint becomes $g(\mathbf{x}) \leq 0$.
d. Define $g(\mathbf{x}) = -f(\mathbf{x})$, the constraint becomes $g(\mathbf{x}) \leq 0$.
e. Define $g(\mathbf{x}) = d - r(\mathbf{x})$, the constraint becomes $g(\mathbf{x}) < 0$.
f. Define $g(\mathbf{x}) = r(\mathbf{x}) - z(\mathbf{x}) - d$, the constraint becomes $g(\mathbf{x}) < 0$.

5.1.2.1 Further Classification of Constraints

The type of constraints, many times, determines the methods to be used to solve an optimization problem. It is generally the case that objectives are user defined or that modifications can be made to them, to make them easier to work with. For example, if the distance between two points is to be minimized, instead of the absolute value function, one can use the Euclidean norm of the distance. This is generally not so straightforward for constraints. Constraints are inherent to the design problem and can very rarely be modified or modifiable. In fact, even if the feasible set (intersection of all the constraints) is a simple convex region, making the determination that it is so is very hard except in simple cases. In light of this, it is generally a good idea to further classify constraints into categories such that it easier to understand what methods can be employed:

1. **Linear and nonlinear constraints:** This classification depends on whether the constraint functions $h_i(\mathbf{x})$ and $g_i(\mathbf{x})$ are themselves linear or nonlinear functions.
2. **Convex and nonconvex constraints:** When $g_i(\mathbf{x})$'s are convex functions, they define a convex feasible set. Convex feasible sets are very desirable in optimization, especially when the objective function is also convex.
3. **Integer constraints:** These constraints limit the values of variables or some function of them to be integers. Integer constraints make design problems very hard to solve; heuristics are often used when they are present.

5.2 Classification of Optimization Problems

Depending on the properties of the objective functions and constraints, optimization problems themselves can be classified into different types. While there is significant overlap between them, the classification helps in choosing what algorithms can and should be applied.

5.2.1 Linear Programs

When optimization problems involve only linear objectives and linear constraints, they are called linear programs (LPs). Linear programs can be considered the simplest form of optimization problems and, therefore, have been studied extensively. As a result, linearization of many large-scale problems is sometimes performed to leverage the vast amount of technical advancements made in solving them. Airline scheduling problems, facility planning, and layout optimization problems are usually studied in the context of linear programs. Linear programs in two dimensions (involving only two design variables) can be solved graphically for a reasonable number of constraints. When more than two variables are involved, one needs an algorithm that can systematically approach the solution. For linear programs, the constraints individually form half spaces, the intersection of which forms what is called a polytope (imagine a polygon in higher dimensions). It is important to know that the optimal solution of a linear program always lies on one of the corners of the polytope.

Before one makes an attempt to solve linear programs, it is important to realize that an incorrectly formulated linear program can be unbounded (the objective can be increased or decreased indefinitely) or simply infeasible (no feasible solutions exist). Figure 5.2 shows two graphical examples. In the left figure, three linear constraints define an unbounded feasible region in the

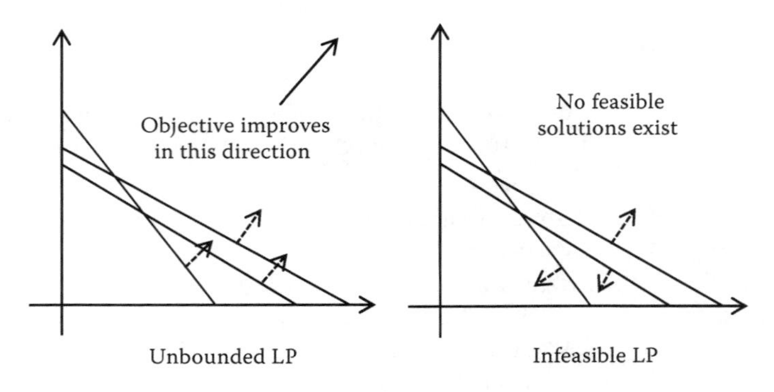

FIGURE 5.2
Graphical examples of unbounded and infeasible linear programs involving inequality constraints. The dotted arrows show the direction of feasibility of the constraints.

top right. If the objective function also improves in this general direction, it can be improved indefinitely, resulting in an unbounded solution. In the right figure, looking at the direction of feasibility of constraints, it is not possible to find a single feasible solution in their region of intersection. Clearly, before one attempts to solve a linear program, one should verify that it has a feasible solution and the solution is not unbounded. We will present methods to do these while describing the simplex algorithm.

There are many methods available to solve linear programs. In this book we will discuss the simplex method and also briefly touch upon interior point methods.

5.2.1.1 Simplex Method

The simplex method was proposed by Dantzig (1963) and exploits the property of linear programs that the optimal solution lies on one of the vertices of the polytope defined by the constraints. This is an extremely important piece of information because it reduces the possibly infinite number of solutions to a finite (but still potentially large) number. Consider the following linear program, which is considered the standard form in the context of the simplex algorithm:

$$\begin{aligned}
\text{Maximize } & c_1x_1 + c_2x_2 + \ldots + c_nx_n \\
\text{subject to } & a_{11}x_1 + a_{12}x_2 + \ldots + c_{1n}x_n = b_1 \\
& a_{21}x_1 + a_{22}x_2 + \ldots + c_{2n}x_n = b_2 \\
& \vdots \\
& a_{m1}x_1 + a_{m2}x_2 + \ldots + c_{mn}x_n = b_m \\
& x_i \geq 0 \quad \forall i \in \{1, \ldots, m\}
\end{aligned} \tag{5.2}$$

Notice that in the standard form, we consider only equality constraints. It does not limit us in any way because inequality constraints can be converted to equality constraints by adding or subtracting positive *slack* variables. The solution method proceeds by also introducing a special variable z, which equals the objective. We will show now the working of the simplex algorithm using an example. Consider the following optimization problem:

$$\begin{aligned}
\text{Maximize } & z = 2x_1 + x_2 + x_3 \\
\text{subject to } & x_1 + 2x_2 + x_3 \leq 8 \\
& 2x_1 + 2x_2 + 5x_3 \leq 30 \\
& 3x_1 + x_2 + x_3 \leq 15 \\
& x_1, x_2, x_3 \geq 0
\end{aligned}$$

TABLE 5.1

Simplex Tableau

Row									Basic Variables
0	z	$-2x_1$	$-x_2$	$-x_3$				$=0$	$z=0$
1		x_1	$+2x_2$	$+x_3$	$+s_1$			$=8$	$s_1=8$
2		$2x_1$	$+2x_2$	$+5x_3$		$+s_2$		$=30$	$s_2=30$
3		$3x_1$	$+x_2$	$+x_3$			$+s_3$	$=15$	$s_3=15$

We first introduce the slack variables and write the optimization problem in tableau form (Table 5.1). Notice that the slack variables have been included as s's. Some authors write them as x's, but it does not affect the treatment of the topic.

5.2.1.1.1 Basic Solution

A basic solution of the linear program is given by first setting $n-m$ variables equal to zero and then solving the constraint equations for the remaining m variables. The simplex algorithm requires that we start with a basic feasible solution (BFS). The variables that are set to zero are called the nonbasic variables (NBVs), while the remaining variables that we solve for are called basic variables (BVs). For the above problem, it is easy to see that the slack variables can be basic variables, while the x's are nonbasic. Setting the nonbasic solutions equal to 0 makes the slack variables s_1, s_2, and s_3 equal to 8, 30, and 15, respectively. Notice that when all the original variables and the slack variables are nonnegative, the solution is feasible; as a result, we have found a basic feasible solution by simple observation.

5.2.1.1.2 Optimality

The next step in the simplex algorithm is to check to see if the current basic feasible solution is also optimal. Recall that we set the x's equal to 0. A look at row 0 of the simplex tableau (Table 5.1) reveals that the solution is not optimal. In fact, all the variables, when increased, will increase the value of the objective. However, we want to make sure that not only does the variable we select have the maximum effect, but also that we increase it in such a way that it does not make the solution infeasible. We choose x_1 because it has the highest coefficient in the objective function. It becomes our *entering variable* in the basic solution. Notice that we cannot increase the value of x_1 indefinitely because it will violate one of the constraints. We see that row 1 limits x_1 to 8, row 2 limits x_1 to 15, and row 3 limits x_1 to 5. Now we need to determine which variable should *leave* the basic solution. Since row 3 limits the value of x_1 the most, s_3 becomes the leaving variable. We perform elementary row operations to make the coefficient of x_1 equal to 1 in row 3 and zero in other rows. The updated tableau is shown in Table 5.2.

TABLE 5.2

Simplex Tableau

Row								Basic Variables
0	z	$-\frac{1}{3}x_2$	$-\frac{1}{3}x_3$			$+\frac{2}{3}s_3$	$=10$	$z=10$
1		$+\frac{5}{3}x_2$	$+\frac{2}{3}x_3$	$+s_1$		$-\frac{1}{3}s_3$	$=3$	$s_1=3$
2		$+\frac{4}{3}x_2$	$+\frac{13}{3}x_3$		$+s_2$	$-\frac{2}{3}s_3$	$=20$	$s_2=20$
3	x_1	$+\frac{1}{3}x_2$	$+\frac{1}{3}x_3$			$+\frac{1}{3}s_3$	$=5$	$x_1=5$

We see that the objective value has increased and the constraint slacks have decreased. The current solution is still not optimal, though, because x_2 and x_3 can be increased further. The method is similar to before, where we select an entering variable and a leaving variable. At each iteration we check for optimality. The optimization process terminates when the objective cannot be increased any further. The optimal solution for the objective problem is $\{3.77, 0.43, 3.26\}$, with the objective value of 11.23. The reader can verify this solution by completing the simplex steps.

It is relatively easy to determine whether a linear program is unbounded using the simplex algorithm. Any time we select a variable as the entering variable, it has a negative sign in row 0. The variable should also have a nonnegative coefficient in at least one of the constraints; if this is not true, the solution is unbounded. It implies that the objective can indefinitely be improved without affecting feasibility. A linear program is infeasible if no solutions exist that satisfy all the constraints. If any of the slack variables end up being negative, the linear program has no feasible solutions. In practical applications, the simplex method is solved using modifications such as the Big M method or the two-phase simplex. The reader is referred to operations research texts for detailed expositions of the topic.

5.2.1.2 Interior Point Methods

In 1984, Narendra Karmarkar proposed an algorithm that, unlike the simplex algorithm, starts from an initial point in the interior of the feasible polytope and recursively approaches the optimum. The main advantage of the algorithm is that it runs in polynomial time, which means that the solution time only increases as a polynomial function of the length of the input. The simplex algorithm, by comparison, is an exponential time algorithm. Large-scale problems such as airline scheduling problems involving thousands of variables and constraints have been successfully solved using the algorithm. Interior point methods are not limited to linear programs. Many convex optimization problems can be solved using interior point methods. Interested readers are referred to Boyd and Vandenberghe (2004).

5.2.2 Integer Programming

Many engineering design problems involve variables that are integers, for example, number of rivets in a sheet metal joint, number of truss members in a structure. Part standardization also discretizes dimensions and strength of components. Integer programs are optimization problems that restrict the variables or some function of them to be an integer. Integer programs are notoriously hard to solve because properties of continuity and differentiability do not readily apply. Exhaustive enumeration of all the solutions is possible for only simpler problems. Integer programs can sometimes be solved by what is known as continuous relaxation; that is, the variables are allowed to be real numbers during the solution process. After a solution is found, it is rounded up or down depending on the type of integer constraint. If continuous relaxation is not an option, heuristics must be used. Branch-and-bound algorithm is a well-known method of solving integer programs. Genetic algorithms are another heuristic that have been shown to be effective in solving integer programs. We cover genetic algorithms later in the chapter.

Example 5.2

Solve the following optimization problem using continuous relaxation:

$$\begin{aligned} &\text{Minimize } f(x) = 3x^2 + 5x - 5 \\ &\text{subject to } 5x \in I \end{aligned}$$

SOLUTION

We first assume that x can take any real value. The minimum of the function lies at the value of x given by

$$x = \frac{-5}{2 \times 3} = -0.833$$

But we know that $5x$ must be an integer, which implies that the decimal part of x can take only values of 0.0, 0.2, 0.4, 0.6, and 0.8. We find the closest value near our continuous solution that will satisfy this requirement, which is 0.8.

Example 5.3

Comment on the solvability of the following problem using enumeration:

$$\begin{aligned} &\text{Minimize } f(\mathbf{x}), \mathbf{x} = (x_1, x_2, \ldots, x_{40})^T \\ &\text{Subject to } x_i \in \{a_1, a_2, \ldots, a_6\} \end{aligned}$$

SOLUTION

In words, the above function asks for the minimum value of $f(\mathbf{x})$ by choosing the values of 40 decision variables, each of which can take six discrete values. Clearly, $6^{40} = 1.34 \times 10^{31}$ different solutions are possible, and their corresponding objective values must be compared to find the best solution. If an enumeration takes 1 nanosecond (10^{-9} seconds), the time it will take to just enumerate the solutions is 4×10^{14} years. This is more than the estimated age of the universe! Problems of this magnitude are routinely encountered in engineering. Fortunately, many times objective functions give us properties that can be exploited so that not all the solutions need to be considered (e.g., the objective function in the previous example was differentiable and allowed for continuous relaxation). If this is not the case, heuristics must be used.

5.2.3 Convex and Nonconvex Programming

Optimization problems can be solved very efficiently if certain convexity conditions are satisfied. By convention, objectives are minimized in optimization problems. If an objective is to be maximized, we can minimize the negative of it and proceed. If, under minimization, the objective of a problem is a convex function of the decision variables *and* the constraints form a convex feasible set, then the problem is called a convex optimization problem. Convex problems are special because a local optimum is also guaranteed to be the global optimum. In other words, if a solution is found where the function attains the minimum value compared to points in the vicinity, it is not only a local minimum but also a global minimum. The theory of solving convex problems is considered almost as complete as that of linear programs (Boyd and Vandenberghe, 2004).

It is generally very hard to determine whether or not a problem is convex. The second derivative test for convexity can be applied only for simple differentiable functions, while finding intersections of constraint sets is a difficult process. If all the constraint sets individually are convex, then their intersection is also going to be convex. We cover these ideas in detail in Section 5.3.1.1.

5.2.4 Mixed Problems

Some optimization problems do not fall under any of the generally described categories. These are usually termed mixed problems. None of the available methods readily apply to these problems. Many times problem-specific simplifying assumptions are made. Heuristics are also routinely employed. Often, preprocessing reveals that modification of the problem might make it tractable. A class of such problems is mixed-integer nonlinear programs (MINLPs). These problems involve nonlinear objectives or constraints, as well as discrete decision variables.

5.2.5 Multiattribute Problems

Many times the utility functions for the decision makers are not explicitly known. Furthermore, sometimes multiple decision makers are involved, and each wants to be able to make a decision optimal for himself or herself. An example of where this is encountered is mass customization problems where a product is to be manufactured with modifications customized for different types of customers. A common methodology employed in such problems is to first find a Pareto front over the attributes, and then allow the end user to choose a design from it. We will present two approaches for solving multiattribute problems at the end of this chapter.

5.3 Optimization Methods

In this section we will go over different methods used for optimization. We reiterate that because of the scope of this book, only an overview is provided. It should be enough for the reader to feel comfortable with how optimization aids engineering design.

5.3.1 Classical Optimization

The term *classical optimization* is used for methods that rely on gradients or approximations of it to find the optimal solution. Classical optimization methods, as a result, require the objective functions or the constraints to satisfy certain properties such as differentiability or convexity. Under strict conditions, classical problems can guarantee convergence to at least a local optimum.

5.3.1.1 Role of Convexity in Optimization

Convexity of the objective function and those of the constraints play an integral role in optimization. As a basic rule, convex problems are easier to solve than nonconvex problems. A convex problem is defined as one where a convex function is minimized over a convex set. In the next few paragraphs an overview of convex sets and convex functions is provided.

Mathematically, a set C can be defined as convex if for any two points $\mathbf{x}$ and $\mathbf{y}$ in C, their convex combination $\mathbf{z}$ (as defined below) also lies within the set.

$$\mathbf{z} = \alpha\mathbf{x} + (1-\alpha)\mathbf{y} \in C \qquad \forall\alpha \in [0,1] \tag{5.3}$$

In two dimensions, this is easy to visualize. From Figure 5.3 we can easily see that a line joining any two points in a convex set will be entirely

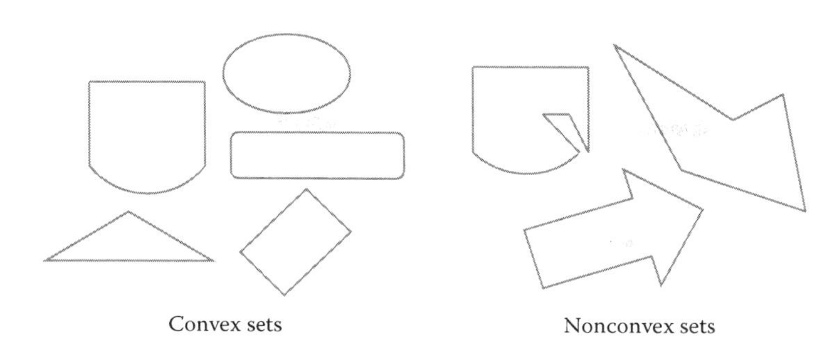

FIGURE 5.3
Examples of convex and nonconvex sets in two dimensions. For any two points within a convex set, all the points on the line joining them are also in the set; this is not true for nonconvex sets.

contained in the set, while this is not true for nonconvex sets. In higher dimensions, visualizing convexity is hard, and therefore the test shown in Equation 5.3 can used.

5.3.1.1.1 Intersection of Convex Sets

The intersection of a collection of convex sets is a convex set. Proving this is pretty straightforward. Consider two sets C_1 and C_2 that are convex and have a nonempty* intersection, that is, $C_1 \cap C_2 \neq \{\phi\}$. Now consider two points **x** and **y** in $C_1 \cap C_2$. It follows that both **x** and **y** are in C_1, and therefore all the points on the line joining them are in C_1 due to the convexity of C_1. The points **x** and **y** are also in C_2, and therefore all the points on the line joining them are in C_2, from the convexity of C_2. From the definition of the intersection of the two sets, we realize that all the points on the line joining **x** and **y** are in $C_1 \cap C_2$, and therefore it is convex. Since the resulting set (intersection) is convex, the result can be extended to a collection of as many sets as we like, so long as they are all convex. Half spaces, balls, ellipsoids, boxes, polyhedra, and their intersections are all convex.

Intersection of a nonconvex set with any other set can be convex or nonconvex, depending on the common points in the two sets. This can be understood using the examples in Figure 5.4.

A union of sets is almost never convex. We will let the reader verify that this is true using different combinations of sets (convex or nonconvex sets). The basic intuition that helps in understanding this is that union involves addition of points to an initial set. These new points may not share the same relationship with the original set.

* The null set is by definition convex, and so is a set with only one element. For technical rigor, a convex set can be defined as a set that does not contain two points such that all the points on the line joining them are not in the set.

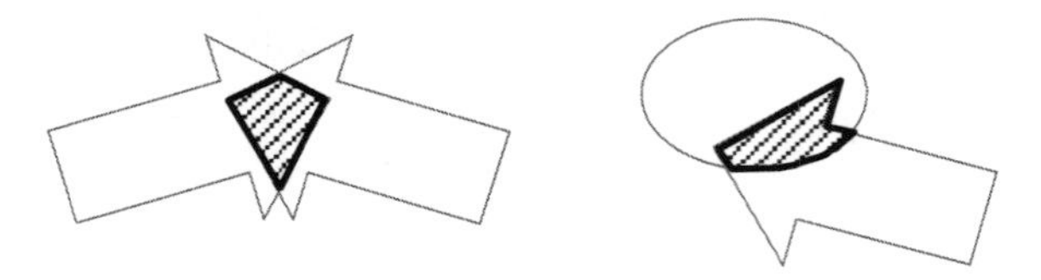

FIGURE 5.4
A nonconvex set can have convex or nonconvex intersections, depending on common points shared with the intersecting set.

5.3.1.1.2 Convex Functions

A function f is convex if a line joining any two points on its graph stays above the function at all times (see Figure 5.5). Mathematically, this implies that for all $x, y \in dom f$ and for all $\theta \in [0,1]$,

$$f(\theta\mathbf{x} + (1-\theta)\mathbf{y}) \leq \theta f(\mathbf{x}) + (1-\theta) f(\mathbf{y}) \tag{5.4}$$

For differentiable functions, this is the same as saying that the second derivative is greater than zero, that is,

$$\frac{d^2 f}{dx^2} \geq 0 \tag{5.5}$$

Also, functions defined over many arguments are convex if their Hessian is positive semidefinite. This can be compactly written as $\nabla^2 f \geq \mathbf{0}$, where the inequality is in the matrix sense; that is, the matrix on the left-hand side is positive semidefinite. A function does not have to be differentiable to be convex. Again referring to the examples of Figure 5.5, the differentiable and nondifferentiable functions are both convex.

Another way to identify a convex function that does not require differentiability and brings out the connection between convex functions and

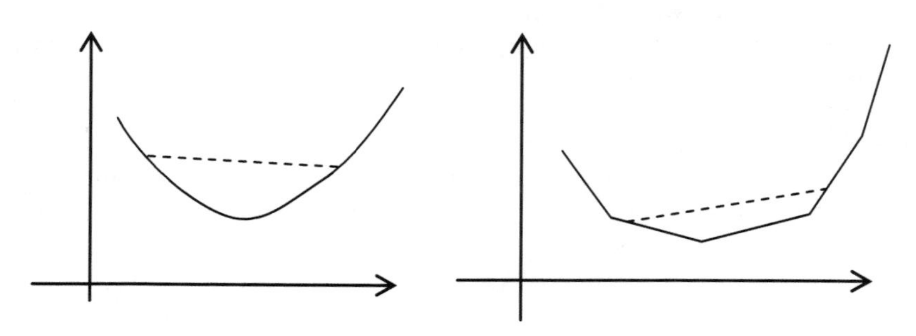

FIGURE 5.5
Examples of convex functions. The function on the left is differentiable, while the one on the right is not.

convex sets is to see if its epigraph is a convex set. The epigraph of a function $f : R^n \rightarrow R$ is the set of all the points in R^{n+1} that lie above the function. Mathematically,

$$\text{epi} f = \{(x, c) \mid x \in \text{dom} f, f(x) \leq c\} \tag{5.6}$$

5.3.1.1.3 Guaranteeing Global Optimality

One of the main reasons for so much interest in convexity is that in convex optimization problems, a local optimum is a global optimum. Linear programs (which are also convex programs) also provide global optimality, but they are very restricted in the kinds of objectives and constraints. To prove that a local optimum is also the global optimum for a convex problem, we start with $\mathbf{x}^*$, which we assume to be the local optimum for the problem. Similarly, assume that $\mathbf{x}_g^*$ is the global optimum. The objective function values at these points are equal to $f(\mathbf{x}^*)$ and $f(\mathbf{x}_g^*)$ and $f(\mathbf{x}_g^*) < f(\mathbf{x}^*)$. Now from the convexity of the feasible set, all **y**'s defined by the following equation are feasible:

$$\mathbf{y} = \alpha \mathbf{x}^* + (1 - \alpha)\mathbf{x}_g^* \quad \forall \alpha \in [0, 1] \tag{5.7}$$

Furthermore, from the convexity of the objective function, we have that function values at **y** are either lower than or equal to either $f(\mathbf{x}^*)$ or $f(\mathbf{x}_g^*)$, or both. This is a contradiction, since $\mathbf{x}_g^*$ was assumed to be a global minimum and $\mathbf{x}^*$ a local minimum (points around $\mathbf{x}^*$ are worse than $\mathbf{x}^*$). Therefore, it must be that $\mathbf{x}^*$ is a global minimum and $f(\mathbf{x}_g^*) = f(\mathbf{x}^*)$.

Example 5.4

Determine if the following optimization problems are convex:

1. Maximize $f(x) = 2 - x^2$, subject to $x^2 - 5 \leq 0$ and $1 - x \leq 0$.
2. Minimize $f(x, y) = 3x^2 + y^2 - xy$, subject to $x, y > -10$.

SOLUTION

1. We can modify the problem as

$$\text{Minimize } \hat{f}(x) = x^2 - 2$$
$$\text{subject to } x^2 - 5 \leq 0 \text{ and } 1 - x \leq 10$$

Since $\frac{d^2\hat{f}(x)}{dx^2} = 2 > 0$, the objective function is globally convex. The constraint $x^2 - 5 \leq 0$ defines the interval $\left[-\sqrt{5}, \sqrt{5}\right]$, and the constraint $1 - x \leq 0$ defines the interval $[1, \infty)$. The feasible set is

$\left[-\sqrt{5},\sqrt{5}\right]\cap[1,\infty)$, which can be rewritten as the interval $\left[1,\sqrt{5}\right]$. This implies that the problem requires minimization of a convex function over a convex set and is therefore a convex optimization problem.

2. Since f is a function of more than one variable, we need to check the Hessian of the function for positive definiteness to determine convexity. The Hessian is given by

$$\begin{bmatrix} 6 & -1 \\ -1 & 2 \end{bmatrix}$$

which is positive definite. Since the feasible set is also a convex set, the problem is a convex optimization problem.

5.3.1.2 Utility Maximization as a Convex Optimization Problem

In Chapter 3, we learned that most decision makers are risk-averse and that a concave function models risk-averse behavior. Under minimization, this utility function becomes convex. If the attributes are defined over a convex set also, then utility maximization becomes a convex optimization problem. This is a very important observation in the context of decision based design. Indeed, the engineering constraints cannot be expected to be convex, but recall that utility functions are defined over the attributes and rarely over the raw decision variables. In such cases, it may be very helpful to know the large convex regions in the attribute space so that the optimum can be found relatively easily and with provable convergence criteria.

5.3.2 KKT Conditions

Karush–Kuhn–Tucker (KKT) conditions, named after the discoverers, provide the necessary conditions for a solution of an optimization problem to be a local optimum. A local optimum is a point that is the best solution (attains the minimum value of the objective) in its vicinity. More succinctly, any perturbation in the solution from the local optimum will worsen the solution. As previously discussed, for convex problems this guarantees global optimality of the solution. Consider the following optimization problem:

$$\begin{aligned} &\text{Minimize } f(\mathbf{x}) \\ &\text{such that } \mathbf{g}(\mathbf{x}) \le 0,\ \mathbf{h}(\mathbf{x}) = 0 \end{aligned} \tag{5.8}$$

where $\mathbf{g}(\mathbf{x}) = \{g_1(\mathbf{x}),\ldots,g_m(\mathbf{x})\}^T$ is a vector of m inequality constraints and $\mathbf{h}(\mathbf{x}) = \{h_1(\mathbf{x}),\ldots,h_n(\mathbf{x})\}^T$ is a vector of n equality constraints. All the functions, f, g's and h's are assumed differentiable.

The method of Lagrange multipliers can be applied as

$$L(\mathbf{x},\boldsymbol{\lambda},\boldsymbol{\mu}) = f(\mathbf{x}) + \boldsymbol{\lambda}^T\mathbf{h}(\mathbf{x}) + \boldsymbol{\mu}^T\mathbf{g}(\mathbf{x}) \tag{5.9}$$

KKT conditions state that under local optimality, that is, if $\mathbf{x}^*$ is the local optimum of the objective function under the constraints, we have

1. Gradient of the Lagrangian is zero, that is, $\nabla L(\mathbf{x}^*,\boldsymbol{\lambda},\boldsymbol{\mu}) = \mathbf{0}$.
2. Feasibility: $\mathbf{g}(\mathbf{x}^*) \leq \mathbf{0}$ and $\mathbf{h}(\mathbf{x}^*) = \mathbf{0}$.
3. Sign condition: $\boldsymbol{\lambda} \geq \mathbf{0}$.
4. Complementary slackness: $\boldsymbol{\lambda}^T\mathbf{g}(\mathbf{x}^*) = \mathbf{0}$.

The KKT conditions become sufficient conditions if the Hessian of the Lagrangian is positive definite at $\mathbf{x}^*$. Many optimization algorithms actually look for KKT conditions to find local solutions to optimization problems.

Example 5.5

Solve the following optimization problem using the KKT conditions.

$$\begin{aligned} \text{Minimize} \quad & f(x_1,x_2) = x_1^2 + x_2^2 + x_1x_2 \\ \text{subject to} \quad & g_1(x_1,x_2): x_1 + 3x_2 - 20 \leq 0 \\ & g_2(x_1,x_2): x_1 - x_2 - 10 \leq 0 \end{aligned}$$

SOLUTION

We write the Lagrangian as

$$L(f(x_1,x_2)) = x_1^2 + x_2^2 + x_1x_2 + \mu_1(x_1 + 3x_2 - 20) + \mu_2(x_1 - x_2 - 10)$$

Writing the KKT conditions, we have

$$\begin{aligned} 2x_1 + x_2 + \mu_1 + \mu_2 &= 0 \\ x_1 + 2x_2 + 3\mu_1 - \mu_2 &= 0 \\ x_1 + 3x_2 - 20 &\leq 0 \\ x_1 - x_2 - 10 &\leq 0 \end{aligned}$$

Now we can enumerate the possible values that the Lagrange multipliers can take.

Case 1: $\mu_1 \neq 0, \mu_2 = 0$, that is, g_1 is active. Therefore, we have three equations:

$$\begin{aligned} 2x_1 + x_2 + \mu_1 &= 0 \\ x_1 + 2x_2 + 3\mu_1 &= 0 \\ x_1 + 3x_2 &= 20 \end{aligned}$$

Solving these equations simultaneously, we get $x_1 = -\frac{10}{7}, x_2 = \frac{50}{7}$. The objective value is 42.85. This is a valid solution because the solution satisfies g_2 as well. However, we cannot be sure that it is the global optimum, so we have to look at other cases.

Case 2: $\mu_2 \neq 0, \mu_1 = 0$, that is, g_2 is active. Therefore, we have three equations:

$$2x_1 + x_2 + \mu_2 = 0$$
$$x_1 + 2x_2 - \mu_2 = 0$$
$$x_1 - x_2 - 10 = 0$$

Solving these together, we get the solution $x_1 = 5, \quad x_2 = -5$, which gives the objective value of 25. The solution also satisfies g_1 and is therefore acceptable. Furthermore, the objective value is less than we found in the previous case, so this solution should be (at least for now) preferred.

Case 3: $\mu_1 = 0, \mu_2 = 0$, that is, both g_1 and g_2 are inactive. Therefore, we have the following equations:

$$2x_1 + x_2 = 0$$
$$x_1 + 2x_2 = 0$$
$$x_1 + 3x_2 - 20 \leq 0$$
$$x_1 - x_2 - 10 \leq 0$$

Solving these equalities and inequalities, we get $x_1 = 0, x_2 = 0$, which gives the objective value of 0 (best found so far).

Case 4: $\mu_1 \neq 0, \mu_2 \neq 0$, i.e., both g_1 and g_2 are active. Therefore, solving the constraint equations only, we get $x_1 = 12.5, x_2 = 2.5$, which gives the objective value of 193.75.

Therefore, the best solution we found so far is when both the variables are equal to 0. To verify that the solution indeed minimizes the function, we check the Hessian of the Lagrangian for positive definiteness. The Hessian is given by

$$H = \begin{bmatrix} 2 & 1 \\ 1 & 2 \end{bmatrix}$$

which is positive definite, verifying that the solution found is indeed the local minimum.

5.3.3 Numerical Optimization Algorithms

As problems become more complex, analyses similar to those in the previous section cannot always be performed. Numerical methods then become more effective in optimization, particularly when looking for an algorithm to implement on the computer. Numerical methods have a local perspective in that they start with a feasible point and progressively move toward the optimum using some iterative algorithm. They still use the concepts of classical optimization, but the search direction from every point along the trajectory is approximated using a variety of rules (some of which may be heuristic). The iterative methods follow a basic scheme as follows:

$$\mathbf{x}_{k+1} = \mathbf{x}_k + \alpha_k \mathbf{x}_k^s \tag{5.10}$$

In words, the above equation says that the next point in the solution trajectory is found by adding to the current point (starting with a random feasible point) a vector given by a search direction $\mathbf{x}_k^s$ times a scaling factor (step size) given by α_k. Notice that both the search direction and the step size are dependent on k; that is, they are allowed to change with every iteration. This is done to accommodate the changes in how quickly the function falls and in what directions. Not all optimization problems are the same; therefore, the scheme used in updating the values of the search direction and scaling factor depends on the algorithm chosen and the problem at hand. Most proposed updating schemes can be shown to work (converge to a local optimum) if the objective function and the feasible set satisfy certain properties, such as convexity, strong convexity, and Lipschitz continuity. Mathematical analyses leading to such proofs are called proofs of convergence. These proofs go a long way in determining the workings of the algorithms. A proof of convergence is not a guarantee, however, that the method is practically applicable.

One condition that the search direction should satisfy is that it should decrease the function. Clearly, the gradient of the function should be relevant here. It turns out that the search directions should make an acute angle with the gradient of the function at $\mathbf{x}_k$. Mathematically,

$$\mathbf{x}_k^{sT} \nabla f_k < 0 \tag{5.11}$$

where ∇f_k is the gradient of the function at $\mathbf{x}_k$. If the above condition is satisfied, the direction is called the *descent direction.* Since gradient is so important in line search methods, we will discuss now how to approximate it.

5.3.3.1 Finite Differencing for Finding Derivative Values

Finite differencing is a numerical way of finding the derivative of a function when a closed-form expression for the derivative is not available. This happens when the function is not differentiable, its closed-form expression is not available, or the function is complicated and it is just easier to find it numerically. The method simply uses the definition of the derivative:

$$\frac{df}{dx} \approx \frac{\Delta f}{\Delta x} = \frac{f(x+\Delta x) - f(x)}{\Delta x} \tag{5.12}$$

Care should be taken to make sure that Δx is small enough; a large value will give the average derivative and may be misleading. At the same time, it should not be so small that machine errors become dominant. For functions of multiple variables, the gradient can be found along each dimension and then arranged in a vector. Notice that the above equation gives the forward difference; that is, it is the gradient when one moves in the positive direction, and therefore it is a skewed approximation of the derivative at x. One could also define the central derivative, by moving incrementally on both sides of x, as

$$\frac{df}{dx} \approx \frac{\Delta f}{\Delta x} = \frac{f(x+\Delta x) - f(x-\Delta x)}{2\Delta x} \tag{5.13}$$

Along the same lines, the second derivative can be approximated as

$$\frac{d^2 f}{dx^2} \approx \frac{f(x+\Delta x) - 2f(x) + f(x-\Delta x)}{\Delta x^2} \tag{5.14}$$

A seemingly obvious way to find the search direction is to simply use the negative of the gradient suitably scaled with a step length. Sometimes practical issues preclude such a simplistic method, and refinements must be made. Interested readers are referred to Boyd and Vandenberghe (2004) for a detailed exposition of the topic.

5.3.4 Heuristics

Heuristics in the optimization context are methods that search for the optimum using "soft" rules that are based on prior experience or simply plausible conjecture. As such, local or global optimality cannot be guaranteed. Heuristics bring tremendous value when the optimization problem does not provide any exploitable properties such as differentiability or convexity. A vast majority of optimization problems are like that. They are nonconvex or the feasible set is nonconvex. Many times the function is nondifferentiable, as in the case of integer variables or constraints. Further problems arise when the objective function is a black-box type; that is, no known functional forms exist or are known to the designer. An example is a large-scale finite element model for an automobile structure. It is nearly impossible to predict with certainty how a particular force will deform the body of an automobile unless a computationally expensive finite element code is run.

5.3.4.1 Random Search

Random search methods start with a solution or a set of solutions and perturb it to improve on it. The premise behind random search is that good enough solutions can be found by chance if the design space is suitably searched, even though at random. There may be rules to perturbation, such as searching only in the vicinity of the original solution in addition to or in place of random search over the whole design space. Understandably, many implementations are possible:

1. Randomly searching the whole feasible set. This becomes problematic in higher-dimensional problems.
2. Weighting the probability density function of the search locations based on where solutions are likely to be found.
3. Starting with searching the whole design space and then narrowing down to "good areas."
4. Perturbing good solutions to improve upon them.
5. Working with a set of solutions instead of one and performing random search starting with each of them.

The methods discussed below, genetic algorithms and ant colony optimization, are special cases of random search methods.

5.3.4.2 Genetic Algorithms

Engineering problems generally have an innate structure. Loosely speaking, most engineering systems can be broken into subsystems, each of which can be separately optimized. Furthermore, system-level attributes are generally monotonic functions of subsystem attributes. As a result, it makes sense to represent system designs as being composed of simpler subsystem designs. These subsystem designs can then be combined every which way, and the overall system performance measured, hoping for an improvement. This line of thinking forms the basis of comparisons between design optimization and biological reproduction and evolution. It is believed, in the context of genetic algorithms (GAs) at least, that good subsystems can be used to progressively arrive at a good overall system, and "evolution" of such a system using operators such as selection, crossover, and mutation will eventually lead to an optimal design. Genetic algorithms work surprisingly well. This is despite the fact that the underlying arguments made in their working are not mathematically rigorous.

Initially proposed by John Holland (1975), genetic algorithms mimic sexual reproduction seen in living beings. Engineering designs are represented with a (usually binary) string, called a *chromosome*, which encodes the decision variables, usually in an appended form. Each instance of the chromosome is called a solution, since it maps to a unique point in the solution space.* The algorithm starts with a set of many randomly generated

* A GA solution lives in the same space as its actual integer or real-valued counterpart. Each chromosome refers to a unique point in solution space, and each point in the solution space refers to a unique chromosome. Since GAs normally discretize the search space, this mapping is a bijection within discretization error.

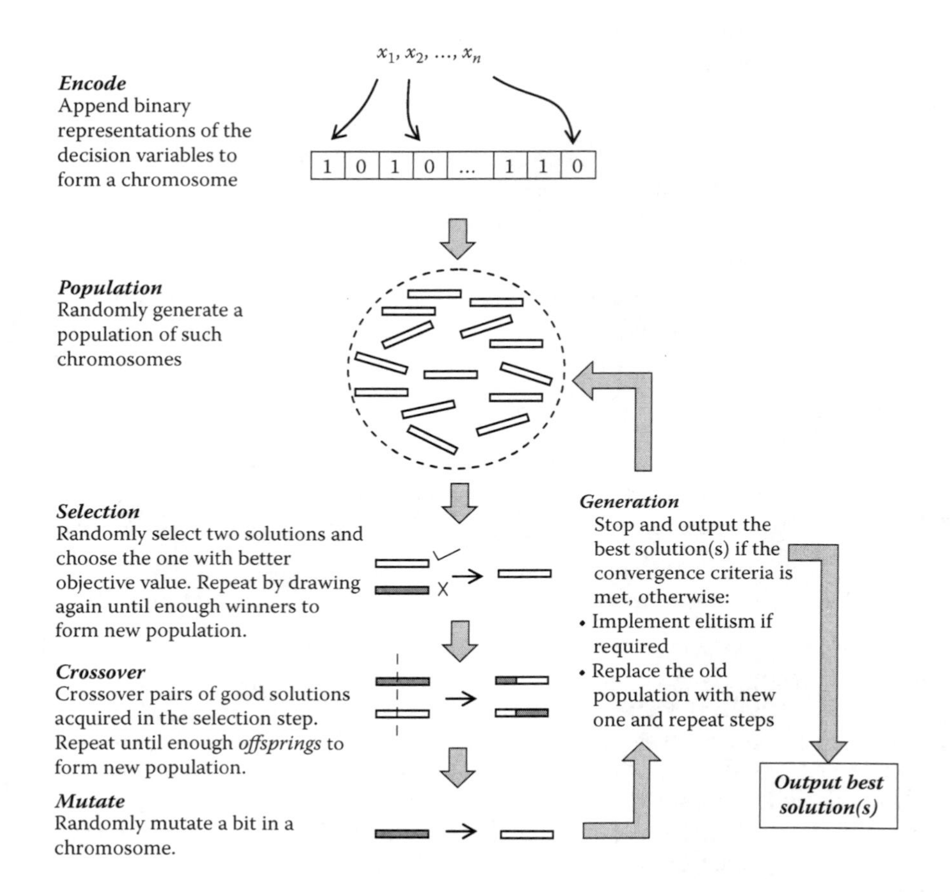

FIGURE 5.6
Schematic of the working of a simple genetic algorithm.

solutions, called the *population*. The size of the population (number of solutions it contains) is called the *population size* and generally remains constant throughout the optimization process. Randomly chosen solutions are then compared based on how good they are, the goodness being determined by the objective value corresponding to a solution. This step is called *selection* (see Figure 5.6). Selection is a way to ensure that good solutions have a better chance to reproduce. After comparison, the winning solution(s) is then chosen for applying the later operators of crossover and mutation. There are many implementations for selection; for example, two or more than two (tournament selection) solutions can be compared together and the best one chosen. Selection can also be done with or without replacement. Depending on the implementation, one increases or decreases the selection pressure; too high a selection pressure and only the best solutions survive. However, this is usually at the cost of diversity in solutions. High selection pressure therefore can get a solution stuck in the local optimum. An example of high selection pressure would be choosing the best one out of a tournament of many (~5) solutions. Only the best solutions at the current stage will reproduce, and other solutions that may have the potential to evolve into much better solutions later will be lost.

Solutions that are selected are then crossed over. Recall that solutions are usually stored as binary strings. Crossover takes two solutions at a time and probabilistically decides on a crossover point, let us say at location m, where $1 \leq m \leq n$ and n is the length of the chromosome. Notice that when $m = n$, no crossover takes place. In a chromosome where the binary representations of design variables are appended, crossover ensures that good portions of one solution have a chance to combine with good portions of another solution. This is the reason selection is performed before crossover. Crossover does not have to happen every time two solutions are selected. In many implementations one specifies a crossover probability up front. Two solutions cross over only with this probability, the rest of the time they are transferred down to mutation unchanged. As with selection, crossover also has many implementations. Instead of one point for crossover, one may choose multiple points on the chromosome for crossover. The benefit of doing this is that solutions evolve quickly; however, on the flipside, crossing over at many points can potentially destroy schemas (building blocks).

The final operator the solution in the generation of a simple genetic algorithm sees is mutation. Mutation is a way to randomly search the solution space. At a probabilistically chosen location on the chromosome, the value of the bit is flipped; that is, if it is 0, it is made 1, and vice versa. In general, mutation perturbs only one variable in the solution, and as a result, it only moves the solution in the vicinity of the already existing solution. Mutation can keep a solution from getting stuck in a local optimum. It is not hard to see that all the main operators—selection, crossover, and mutation—mimic what has happened in the evolutionary history of living beings. Further operators, such as niching (keeping subpopulations of different traits) and elitism (saving good solutions across generations), can be used as well. Interested readers are referred to a textbook on genetic algorithms.

A significant amount of research has gone into understanding the theoretical basis of why genetic algorithms perform so well. The best known of these theories is the schema theorem. In most genetic algorithm implementations, a solution is stored in what is called a chromosome, as an appended list of solution variables containing building blocks (called schemas). The schema theorem shows that the selection operator guarantees that good building blocks get exponentially better representation in the population with each generation. This allows such good building blocks to have an increased chance of combining with other good building blocks to reach the optimum.

TABLE 5.3

Initial Randomly Generated Population

Solution Number	Binary Representation for x					Binary Representation for y					x	y	Objective Value
1	1	0	0	1	0	0	1	0	0	0	0.8065	−2.4194	9.329
2	0	0	1	1	1	0	0	0	1	1	−2.7419	4.0323	7.701
3	1	0	0	1	0	0	0	1	0	1	0.8065	−3.3871	15.728
4	1	0	1	0	0	1	1	0	1	0	1.4516	3.3871	10.614
5	1	1	0	0	1	1	1	0	0	0	3.0645	2.7419	28.895

Example 5.6

Minimize the algebraic function $x^3 + y^2 - xy + 1$ within the intervals $x \in [-5,5]$ and $y \in [-5,5]$.

SOLUTION

To illustrate the working of the GA we use a short chromosome length of 10 bits (5 bits for each variable appended together) and a population size also of 5. Table 5.3 shows the initial randomly generated population. The five bits corresponding to each variable are shown. The decimal value is calculated as

$$x_{decimal} = \text{lower limit} + \frac{\text{decimal representation of the binary string}}{2^{\text{number of bits}} - 1}(\text{upper limit} - \text{lower limit})$$

The value of x in the first row can be calculated as

$$x_{decimal} = -5 + \frac{1 \cdot 2^4 + 0 \cdot 2^3 + 0 \cdot 2^2 + 1 \cdot 2^1 + 0 \cdot 2^0}{2^5 - 1}(5 - (-5)) = 0.8065$$

The crossover probability was set to 0.7, while the mutation probability was at 0.2. Figure 5.7 shows the solution progress with the number of generations. It can be seen that the average fitness moves a little bit, while the best solution improves monotonically.

The GA was allowed to run for about 90 generations, and the optimal solution found is {–5, –2.742} and the objective value is –130.191. The true optimum lies at {–5, –2.5}, with the objective value of –130.25. Notice that the minimum resolution we have for the number of bits we chose is 10/31, or 0.322; therefore, we should expect an error of about half that number.

5.3.4.3 Ant Colony Optimization

Ant colony optimization (ACO) is another heuristic optimization method that tries to mimic biological systems. ACO was proposed in 1992 by Marco Dorigo in his PhD thesis. The method draws from the foraging behavior of a colony of ants that try to find the optimal path from their

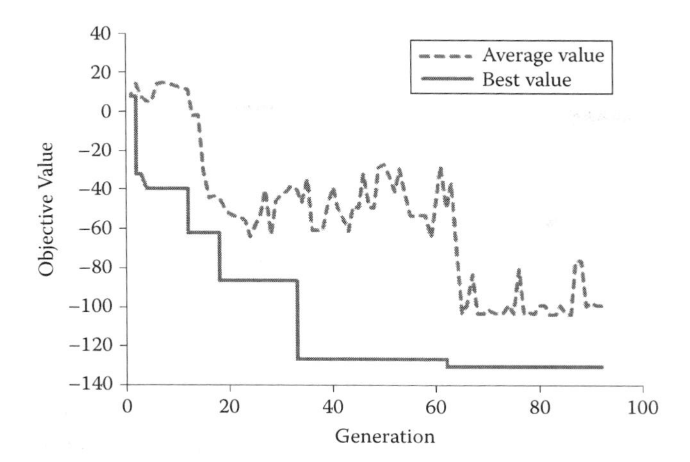

FIGURE 5.7
Solution progress in a genetic algorithm with the number of iterations.

colony to the source of food. The ants are, of course, software constructs in the algorithm implementations. They are relatively simple entities that interact indirectly through what is termed *stigmergy*. An example of this in real-life insects would be chemical pheromones. ACOs are used for combinatorial optimization problems, and they arrive at a solution by incrementally constructing it.

There are three procedures involved in the implementation of an ACO:

1. Construct ant solutions
2. Update pheromones
3. Daemon actions

Many independent agents (ants) explore the search space, which is represented by a fully connected graph.* Each ant moves to the adjacent site randomly or in response to the amount of pheromones present. The length of the path traversed by the ant can be limited using some number based on the properties of the problem. The amount of pheromone deposition is a function of the fitness of the solutions or partial solutions found. Pheromones increase the probability that the next ant(s) will pass through that node. Pheromones also have an evaporation rate associated with them; therefore, less frequented paths eventually lose all pheromones, limiting ants searching there. Ants therefore do not learn themselves, but modify their environment and their emergent behavior in response to it solves the problem. Constraint handling is performed by limiting what nodes ants can or cannot

* A graph is an ordered pair $G = (V, R)$, where V is the nodes and R the edges connecting them.

travel. Sometimes constraints are implemented as soft heuristics by using a penalty for violating them. Any number of heuristics can be implemented in the working of the method.

5.3.4.4 Simulated Annealing

Kirkpatrick, Gelatt, and Vecchi (1983) discuss the strategy mimicking the metallurgical process of annealing in finding a solution to an optimization problem. In annealing, a substance is cooled at a controlled rate that helps in inducing properties such as long-range order, ductility, and malleability. As with the heuristics discussed so far, in simulated annealing, solutions are randomly generated at first. Based on the fitness (value of the objective) and some parameters devised by the implementer, a solution is perturbed to a nearby solution. This nearby solution is sampled and accepted based on some global parameter (similar to temperature in annealing) and its fitness. Simulated annealing is good at finding decent solutions but does not necessarily get the true global optimum. The implicit premise in simulated annealing–based optimization is that the initial population of solutions will not generally be very good. As a result, the amount of allowed perturbation is large, so that the whole feasible space can be searched. As the algorithm progresses, the global "temperature" is decreased, in hopes that good solutions have been found. There is still a possibility to improve the solutions, but the ability to search is limited based on the temperature. In a nutshell, simulated annealing searches for good solutions throughout the design space initially and then limits the search in the vicinity of good solutions later on, to speed up convergence.

5.4 Multiobjective Optimization

So far in this chapter we have dealt with optimization of a single objective. Many times in engineering problems, multiple attributes need to be optimized simultaneously. Indeed, it is possible to combine the attributes into a single attribute, but as we saw in Chapter 3, there is only one way to do it properly, that is, use a utility function. If such a utility function is available, then one would just use the methods discussed earlier in the chapter. Utility functions, however, may not be readily available, or there could be disagreement about which utility function to use when multiple decision makers are involved. In such cases, a unique single solution is not possible or desirable; one has to instead resort to the set of Pareto optimal solutions. We discussed Pareto fronts in Chapter 3, where we argued that they represent unavoidable trade-offs one has to make in a design situation. In this section we will talk about two ways of generating them.

5.4.1 Optimizing Convex Combination of Attributes

Assume that you have n attributes $y_1(\mathbf{x}), y_2(\mathbf{x}), \ldots, y_n(\mathbf{x})$ to be minimized by manipulating the decision variable vector $\mathbf{x}$. A convex combination of the attributes is given by

$$y(\mathbf{x}) = \sum_{i=1}^{n} \alpha_i y_i(\mathbf{x}) \tag{5.15}$$

where

$$\sum_{i=1}^{n} \alpha_i = 1 \quad \text{and} \quad \alpha_i \in [0,1] \quad \forall i$$

The convex combination $y(\mathbf{x})$ is then minimized subject to the constraints of the problem, for a given combination of α_i's. The solution will give us one point on the attribute space. We repeat the optimization process for all the possible combinations of α_i's and plot the solutions in the attribute space. The set of the solutions forms a Pareto front. Ideally, we would want to implement all combination of α_i's, but since optimization problems are generally computationally expensive, the coefficients are varied in a discrete way. Finer discretization can be done in the case where one wants higher resolution in the Pareto front. The issue with this method is that as the number of attributes increases, the number of times the optimization has to be run increases exponentially. Heuristics such as nondominated sorting, which is discussed in the next section, are performed for attributes over many attributes.

Example 5.7

Find the Pareto front for the following optimization problem:

$$\text{Minimize } \{y_1(\mathbf{x}), y_2(\mathbf{x})\} \quad \mathbf{x} = (x_1, x_2)^T$$

$$\text{where} \quad y_1(\mathbf{x}) = x_1^2 - x_2$$

$$y_2(\mathbf{x}) = x_2^2 - x_1$$

$$\text{subject to } x_1 \in [0,10], x_2 \in [0,10]$$

SOLUTION

We write the convex combination of the attributes as

$$y(\mathbf{x}) = \alpha y_1(\mathbf{x}) + \beta y_2(\mathbf{x}) = \alpha x_1^2 + \beta x_2^2 - \alpha x_2 - \beta x_1$$

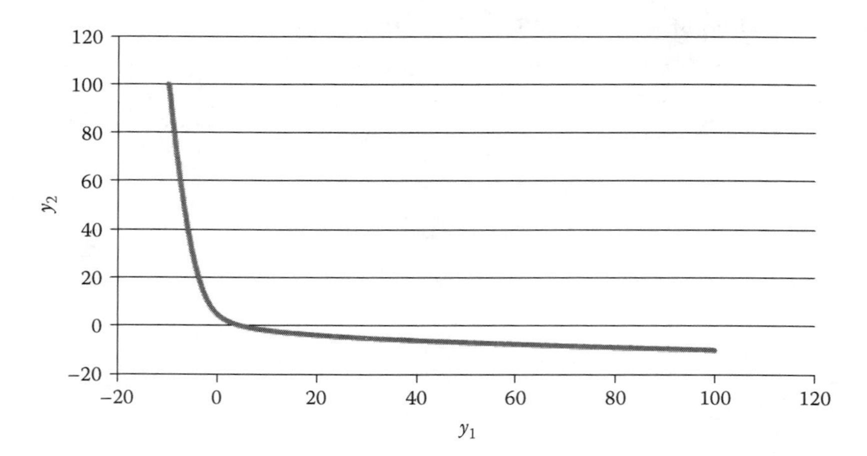

FIGURE 5.8
The Pareto front generated for Problem 5.7 by minimizing the convex combination of attributes (the discrete points are shown joined using a smooth curve).

We find the partial derivatives and equate them to zero:

$$\frac{\partial y(\mathbf{x})}{\partial x_1} = 2\alpha x_1 - \beta = 0$$

and

$$\frac{\partial y(\mathbf{x})}{\partial x_2} = 2\beta x_2 - \alpha = 0$$

Therefore,

$$x_1 = \beta/2\alpha \quad \text{and} \quad x_2 = \alpha/2\beta$$

Figure 5.8 shows the values of the two objectives when α is varied between 0 and 1 while β is equal to $1 - \alpha$.

5.4.2 Nondominated Sorting

The method of nondominated sorting (Goldberg, 1989) involves working with a set of solutions as in a genetic algorithm. The selection step of the GA is modified somewhat since multiple attributes are involved. We first look for dominance between solutions: a solution $\mathbf{x}_1$ dominates a solution $\mathbf{x}_2$ if the following relationship holds between their corresponding attribute values (y's):

$$y_i(\mathbf{x}_1) \leq y_i(\mathbf{x}_2) \quad \forall i \in \{1,\ldots,n\} \tag{5.16}$$

and

$$\exists j \in \{1,\ldots,n\} \quad \text{such that} \quad y_j(\mathbf{x}_1) < y_j(\mathbf{x}_2)$$

Here we have assumed, as previously, that the attributes are minimized. In words, the above relations state that a solution is said to dominate another if it is at least as good as the other solution in all the attributes, and it is strictly better in at least one attribute. Dominance does not generally exist between a pair of solutions, especially in the later generations of the genetic algorithm. For our example, it is possible that the above relationships do not establish that $\mathbf{x}_1$ dominates $\mathbf{x}_2$, or the equally plausible case that $\mathbf{x}_2$ dominates $\mathbf{x}_1$. In such cases the solutions are mutually nondominated. Indeed, it is true that a Pareto front is a collection of nondominated solutions; however, the converse is not true. That is, just because all the solutions in our population are mutually nondominated, we do not necessarily have the Pareto front. We have to be reasonably sure that for the given set of solutions we cannot find *any* other feasible solution that dominates (even one of) them. When such a solution set is found, the solutions fall on the Pareto front. A Pareto front is a set of solutions that are mutually nondominated and there are no feasible solutions that dominate any one of the points on the front. How do we then converge to a Pareto front? Well, at the very least, we will need a way to compare nondominated solutions. We will now present how a well-known algorithm called Nondominated Sorting Genetic Algorithm II (NSGA-II) does this (Deb et al., 2002).

In NSGA-II, the operators of selection (domination), crossover, and mutation are first run to get the second population. The two populations are then combined. Each solution is compared to every other solution in the combined population for dominance, as described above, and scored based on how many solutions it dominates. The solutions are then sorted based on this score. Notice that domination is a transitive property in that if one solution dominates another in a pairwise comparison, it also dominates all the solutions dominated by the latter. As a result, if a solution dominates another, it also ranks higher after sorting (because of a higher score). In the sorted population, multiple solutions may have the same score. The highest scoring solutions form the first (best) nondominated *front*, the next highest scoring the second nondominated front, and so on. Since the population size should remain constant during GA, the solution is truncated at the population size, but all the solutions with the same scores as the truncation point are kept aside for the *crowdism* operator. Within this crowdism set, by definition all the solutions have the same score. To select which to put in the new population and which to discard, the algorithm looks at the normalized distance between them across all the *attributes*. A solution that is close to other solutions (has similar objective values) is *discarded* in favor of a solution that is far away from other solutions. The premise behind doing this is that a well-distributed Pareto front is favorable to a front where all the solutions are clumped together.

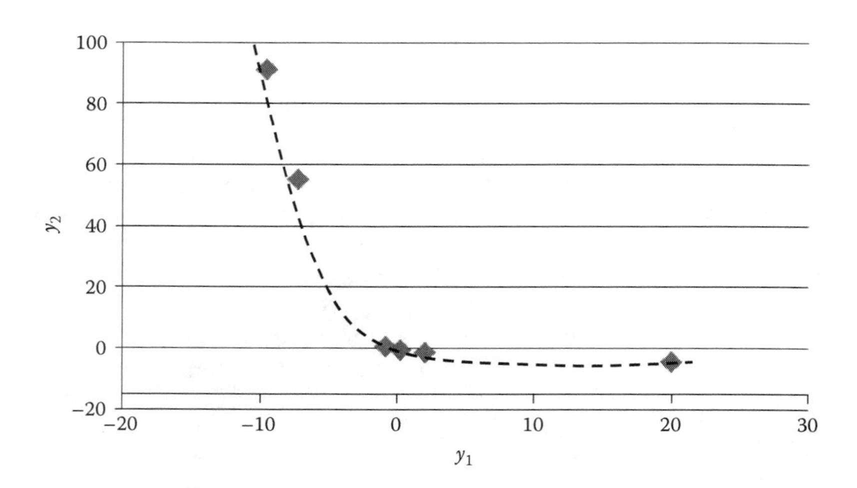

FIGURE 5.9
The Pareto front generated using NSGA-II for Problem 5.7. A small population size was used for only a few generations, which is why the solutions are not well spread out.

It is possible that within a few generations one ends up with a population where no solution dominates another. As we discussed earlier, this does not imply that the true front has been found. NSGA-II is continued until a user-specified criterion is met (e.g., number of generations) or until no further improvement in the front is found for a few generations. Needless to say, finding the Pareto front using domination requires sampling all of the solutions or, if this is not possible, a vast number of them. As a result, finding the front is computationally intensive, usually more so than finding the optimum in single-attribute cases. With well-designed evolutionary algorithms, once convergence criteria are met, one is left with an approximation of the Pareto front that serves the purpose for most cases. Figure 5.9 shows the Pareto front for the optimization problem in Example 5.7 using NSGA-II. Even for a very small population size run for only a few generations, we can see the ability of the algorithm to approximate the Pareto front. NSGA-II with a large population size can easily populate the whole true Pareto front for a problem of this size.

Problems and Exercises

1. In the context of optimization, what is the holy grail of decision based design?

2. Determine if the following optimization problems are convex or nonconvex.
   a. Maximize $f(x) = x^2$, subject to $x - 5 \leq 0$ and $1 - x^2 \leq 0$.
   b. Minimize $f(x, y) = x^2 + y^2 + xy$, subject to $x, y > -56$.
3. What are the advantages of convex functions, convex constraints, and both in an optimization problem?
4. Solve the following linear programs using the simplex method:

a. Minimize $z = x_1 - x_2$

subject to
$$x_1 + x_2 \leq 8$$
$$3x_1 - x_2 \leq 4$$
$$x_1 \leq 3$$
$$x_1, x_2 \geq 0$$

b. Maximize $z = 2x_1 + x_2 + x_4$

subject to
$$x_1 + x_3 \leq 8$$
$$x_4 \leq 6$$
$$x_1 + x_2 + x_3 \leq 15$$
$$x_3 - x_4 \leq 3$$
$$x_1, x_2, x_3, x_4 \geq 0$$

5. What are interior point methods in optimization?
6. Solve the following optimization problems using KKT conditions:

a. Minimize $f(x_1, x_2) = x_1^2 + x_2^2 - 3x_1$

subject to
$$g_1(x_1, x_2): x_1 + x_2 - 5 \leq 0$$
$$g_2(x_1, x_2): x_1 - 3x_2 - 8 \leq 0$$

b. Minimize $f(x_1, x_2, x_3) = x_1^2 + x_2^2 + x_3^2$

subject to
$$g_1(x_1, x_2, x_3): x_1 + x_2 + x_3 - 20 \leq 0$$
$$g_2(x_1, x_2, x_3): x_1 - 3x_2 \leq 0$$
$$g_3(x_1, x_2, x_3): -30 - x_3 \leq 0$$

7. Why are linear programs not very relevant in the context of decision based design?
8. What are genetic algorithms? Write the basic intuition behind why genetic algorithms work.
9. What is schema theorem in genetic algorithms? Can it be considered a proof of convergence?
10. What is ant colony optimization? Why do you think it can work better than genetic algorithms for some problems?

11. What is simulated annealing? Explain the basic premise behind simulated annealing.
12. Decide on the correct approach to use for the following optimization problems:
    a. Objective—convex, differentiable; constraints—convex
    b. Objective—linear, differentiable; constraints—linear
    c. Objective—linear, differentiable; constraints—convex
    d. Objective—nondifferentiable, Lipshitz; constraints—convex
    e. Objective—nondifferentiable; constraints—nonconvex, integer
13. Explain the following terms in the context of genetic algorithms:
    a. Chromosome
    b. Population
    c. Population size
    d. Selection
    e. Mutation
    f. Crossover
    g. Generation
    h. Nondomination
14. Discuss multiobjective optimization using a convex combination of attributes. Must we always use a convex combination?
15. Discuss multiobjective optimization using genetic algorithms. What is the benefit of using genetic algorithms for multiobjective problems?
16. What is nondominated sorting? How does NSGA-II work?
17. Solve the following optimization problems using NSGA-II:
    a. Minimize $\{y_1(\mathbf{x}), y_2(\mathbf{x})\}$ $\mathbf{x} = (x_1, x_2)^T$

       where $y_1(\mathbf{x}) = (x_1 - 5)^2 + (x_2 - 5)^2 ; y_2(\mathbf{x}) = (x_1 - 1)^2 + (x_2 - 1)^2$

       subject to $g(x) : (x_1 - 3)^2 + (x_2 - 3)^2 \geq 1$
    b. Minimize $\{y_1(x), y_2(x), y_3(x)\}$

       where $y_1(x) = \sin x; y_2(x) = \cos x; y_3(x) = x$

       subject to $x \in \left[0, \frac{3\pi}{4}\right]$

6

Simulation Methods in Engineering Design

6.1 Introduction

Most real-life situations involve uncertainty, and engineering design is no exception. As we have been doing in this book, we assess and model uncertainty mathematically. We define a physically measurable quantity (or multiple quantities) and use a probability density function (pdf) to assess how its realizations behave. In decision based design, we are eventually concerned with the pdf of the decision maker's utility function so that we can find its expectation. Getting to it, however, requires us to first model the pdf of the independent decision variables (inputs) and then use a model that relates the input to the output of the system to get the pdf of the output variables (attributes). The pdf of the attributes can then be used to either assess the pdf of the utility function or directly find its expectation, as shown in Figure 6.1.

In theory, we can analytically find closed-form expressions for the pdf of the attributes, and consequently the utility. For simpler systems this may be the case, but doing so for most systems is extremely difficult, and sometimes such a method does not even exist. It is therefore easier in most cases to simulate the system using the model of the input–output relationships for each realization of the input variables. This will give us realizations of the attributes. Using a well-suited distribution, we can fit the data derived from the attribute realizations. Alternatively, we could directly find realizations of the utility function, if we have it available, and calculate its expectation by averaging all the values.

6.1.1 Need for Simulating Engineering Systems

A designer must know how an engineering system will behave during use. Instead of performing expensive tests, one can learn this behavior by creating a mathematical model of the system and subjecting it to different operating conditions (values of input variables). It is very critical that this model not only be a good representation of the actual system, but also be subjected to similar (simulated) operating conditions that the actual system would encounter during operation. Of particular interest to designers is the ability

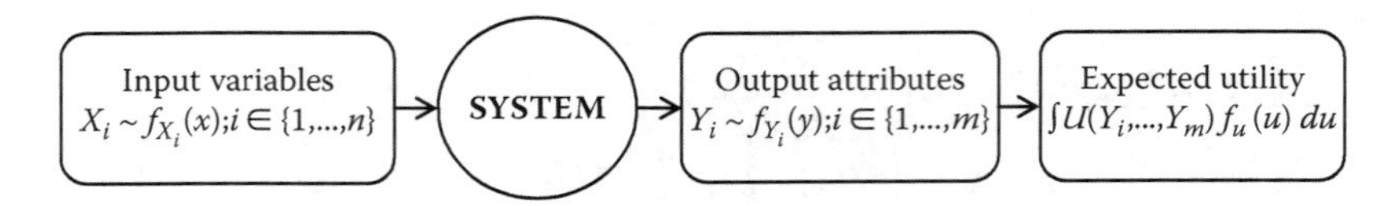

FIGURE 6.1
Schematic of uncertainty propagation in an engineering system.

to understand the effect of uncertainty* on the working of engineering systems, since almost all of them are subjected to uncertainty. Knowing the effect of uncertainty not only helps determine the performance parameters of the system, but also helps model its reliability or the ability to fulfill stated requirements in a probabilistic sense. Of course, the designer needs to know the true nature of uncertainty (what variables are affected and what their pdfs are) to accurately predict the performance of the system. He or she also needs to know how uncertainty propagates through the system and how it eventually affects the expectation of the utility.

6.2 Simulation Approach

Engineering system simulation under uncertainty roughly involves four steps:

1. Identifying sources of uncertainty, that is, variables that are uncertain (called random variables)
2. Modeling the uncertainty in the random variables, that is, fitting distributions to the random variables while also incorporating dependence relationships between them
3. Simulating the engineering system that involves generating realizations of the input random variables and using the input–output relationship of the engineering system to acquire realizations of the outputs (attributes)
4. Finding the utility values for realizations of the attributes and averaging them to find the expected utility

Notice that to simulate a system we would need a way to determine the output of an engineering system given an input. The relationship between inputs and outputs (under uncertainty) is also essential for optimization, as

* System simulation is a broader term that may or may not include uncertainty. In this chapter, we are concerned with only simulation of random design variables and their effect on designer utility.

we have seen in Chapter 5. Finding this relationship is usually termed modeling. Our focus in this chapter is not on this modeling; we assume that such a representation (mathematical or otherwise) exists.

6.2.1 Monte Carlo Simulation

The Monte Carlo simulation method is used to collect realizations of outputs of a system whose mathematical description is available but evaluating closed-form expressions for pdfs of the outputs is prohibitive or even impossible. These realizations can be used to determine many characteristics of the system, such as probability distributions of system outputs and the system's probability of failure, among others. In technical literature, Monte Carlo simulation mostly refers to step 3 above. Use of the Monte Carlo method requires having access to a large number of realizations of input random variables with a given dependence structure. For each realization the output is *deterministically* calculated and its value stored. Once enough realizations of the outputs are available, one can fit a pdf or calculate the probability of their exceeding a critical value, and so forth. The Monte Carlo method can be extremely computationally expensive, and its use is discouraged when closed-form expressions of the output distributions can be evaluated. Unfortunately, a majority of engineering systems are too complex to allow finding such expressions. Work-arounds such as response surface methods and importance sampling are constantly being explored and refined.

6.2.2 Identifying Sources of Uncertainty

Identifying sources of uncertainty is a difficult task. When we specify design parameters, we generally realize that there will be a probability distribution associated with these parameters, when the design is finally implemented. For example, there are manufacturing tolerances involved, cost of raw materials is uncertain, labor costs and overheads cannot be completely determined, and finally, operating conditions for any design are also uncertain. We must account for these uncertainties during design. It is hard to tell which of these sources of uncertainty should be considered when we predict system behavior, or when setting performance targets. To trim down the list of random variables, a variance approach can be used. For each uncertain variable, the upper and lower limits are defined and the system outputs calculated by changing the variables one at a time, while others are kept at their baseline values. The variables with the most effect on the outputs are then selected, and their probability distributions carefully modeled. In Chapter 3, we discussed two types of uncertainty, reducible and irreducible. For reducible uncertainties, value of information studies can be performed, as we saw in Chapter 3. For irreducible uncertainties with substantial effect on output variables, one should devise ways to mitigate them, or if that is not possible, to very accurately assess their probability distributions. For the remainder of this chapter, we assume that which variables' effects we want to consider have been identified.

6.2.3 Modeling the Uncertainty in Random Variables

Modeling the uncertainty in random variables involves finding their probability density functions. Pdfs need to be assessed from data, using expert judgments, or both. In this section we briefly consider the methods for assessing pdfs, given realizations of the random variable (data).

6.2.3.1 Moment Matching Method

Moment matching involves choosing a distribution whose moments (mean, standard deviation, etc.) match the moments of the data. The moment matching method does not focus on the pdf of the data; instead, it relies on the premise that the chosen distribution will be a good type of distribution for the data. Since normal distribution is completely determinable with just mean and standard deviation, it is commonly used to fit data using the moment matching method if it is believed that the data indeed follows a normal distribution.

The general formulation, while trying to fit a distribution $f_X(x,\mathbf{p})$ to given realizations of X where $\mathbf{p}$ is the vector of parameters, is to find $\mathbf{p}$ such that

$$\text{Mean matches: } \mu_X = \int x f_X(x,\mathbf{p})\,dx = \mu_{Sample}$$

$$\text{Variance matches: } \sigma_X^2 = \int (x-\mu_X)^2 f_X(x,\mathbf{p})\,dx = \sigma_{Sample}^2$$

$$\text{Skewness matches: } \gamma_X = \int \left(\frac{x-\mu_X}{\sigma_X}\right)^3 f_X(x,\mathbf{p})\,dx = \gamma_{Sample} \ldots, \text{ and so on}$$

Many times finding $\mathbf{p}$ involves solving a set of simultaneous equations. Each of the expressions above constrains the parameters somewhat, and the eventual goal is to have as many equations as there are parameters. Not all moments need to be matched; depending on the problem, one can choose to fit only the first two or three moments. Alternatively, an optimization problem can be solved that minimizes the error between the sample moments and those of the fitting distribution.

Example 6.1

Find the best-fit beta distribution for the data in Table 6.1.

SOLUTION

A beta distribution is completely determined by the lower and upper limits, a and b, and the shape parameters, α and β. Looking at the data, we can set

TABLE 6.1

Example Data

7.467957	5.073743	3.268359	3.468399
5.124154	5.663047	6.270374	6.495231
6.286403	4.114085	3.123684	3.08637
4.020997	5.660625	7.612657	7.219843
5.001187	6.557459	5.338664	5.829005

the lower and upper limits at 3 and 8, since they include all the realizations. The pdf of the distribution with these upper and lower limits is

$$f_X(x;\alpha,\beta) = \frac{(x-3)^{\alpha-1}(8-x)^{\beta-1}}{B(\alpha,\beta)\cdot 5^{\alpha+\beta-1}}$$

where $B(\alpha,\beta)$ is the beta function. The shape parameters can be determined using the following relationships:

$$\mu_X = \frac{\alpha}{\alpha+\beta}$$

$$\sigma_X^2 = \frac{\alpha\beta}{(\alpha+\beta)^2(\alpha+\beta+1)}$$

The mean of the distribution, once normalized between 0 and 1, is equal to 0.467, and its variance is equal to 0.083. Solving the above equations we get $\alpha = 0.933$ and $\beta = 1.065$. The pdf is therefore given by

$$f_X(x;\alpha,\beta) = \frac{(x-3)^{-0.067}(8-x)^{0.065}}{5.016\cdot B(0.933,1.065)}$$

6.2.3.2 Incorporating Expert Judgments

Sometimes, there is limited data available about a distribution to confidently encode it in a probability distribution. It makes sense then to use expert judgments regarding the distribution that the data follows. The expert opinion is embedded into a probability distribution called the prior. The prior is then combined with the limited data we have to come up with the posterior distribution, using the Bayesian method. By definition, the prior takes into account the initial subjective understanding about the distribution, which is updated as and when more information becomes available. This method of updating is based on Bayes' theorem and is given by

$$f_X'(x|d) = \frac{f_D(d|x)\cdot f_X(x)}{\int_{-\infty}^{\infty} f_D(d|x)\cdot f_X(x)dx} \tag{6.1}$$

Here $f_X(x)$ is the prior distribution and $f'_X(x)$ is the posterior distribution of the variable X, while $f_D(d|x)$ is the likelihood of getting the data D (e.g., n successes out of m trials) given the prior distribution of X. Many times we try to assess the probability distribution of the relative frequency (probability) of an event, for example, the probability of a tornado hitting a town given its path a few days prior. The Bayesian method can help make a decision when data are scarce. On the other hand, if the prior knowledge is incorrect, it can corrupt even good information coming in the form of data. Priors therefore should correctly model the decision maker's belief and be elicited very carefully. Maximum entropy methods, discussed later, can provide a normative way of achieving this.

6.2.3.2.1 Conjugate Priors

Conjugate priors are those that, when updated with a certain specific kind of likelihood function, result in a posterior of the same kind as themselves. Since Bayesian updating routinely requires calculations (numerical or otherwise) of complicated functions, this conjugate prior property is very useful, as it simplifies the assessment of the posterior. One of the most famous conjugate priors is the beta distribution prior when updated with a binomial distribution. Beta distribution priors also provide substantial flexibility in the modeling of random variables and are double bounded like most real-life quantities—hence their ubiquity. Consider the design of a car suspension. Assume that for each design we want to know the probability that it will be able to withstand a 1-mile run in a rough terrain. Since we are unsure of the probability P, we ask an expert to encode his or her assessment of the probability and assume that he or she provides a beta distribution in the form

$$f_P(p) = \frac{p^{\alpha-1}(1-p)^{\beta-1}}{B(\alpha,\beta)} \tag{6.2}$$

The value of the parameters α and β the expert provides show his or her degree of belief. In a beta distribution, mean is given by $\frac{\alpha}{\alpha+\beta}$, while the sum $\alpha + \beta$ is roughly the number of experiments the expert thinks his or her opinion is worth. So if the expert is not very confident, he or she would use low values of α and β, while he or she would use large values if he or she is very confident. In both cases, the expert would make sure that $\frac{\alpha}{\alpha+\beta}$ is equal to his or her assessment of the mean of P. If incoming test data reveal that the suspension withstands the terrain in m cases out of n trials (binomial distribution), the updated distribution will be

$$f_P(p) = \frac{p^{\alpha+m-1}(1-p)^{\beta+n-m-1}}{B(\alpha+m,\beta+n-m)} \tag{6.3}$$

It is clear that no calculations were needed to get this posterior pdf. Having conjugate priors therefore saves us the step of finding the complicated integrals.

6.2.3.3 Maximum Entropy Method

Claude Shannon (1948) in his seminal work defined the concept of entropy, sometimes also referred to as information entropy, associated with a probability distribution. It represents our lack of knowledge, in an average case, about where the realizations of a random variable are likely to fall. For discrete distributions the entropy expression is

$$H = -\sum_{i} p_i \log p_i \tag{6.4}$$

And for continuous distribution, this expression becomes

$$H = -\int f_X(x) \log(f_X(x)) dx \tag{6.5}$$

Jaynes (1957) then proposed that the best distribution to use, under limited information, is the one that maximizes this entropy subject to our current state of information. The intuitive reasoning is that the maximum entropy distribution (MED) maximizes our ignorance or makes the least assumptions over what we already know and, as such, can be deemed normative. There are many readily determinable distributions, depending on what we know. For example, if all we know about a distribution are its upper and lower limits, the MED is the uniform distribution. This can be easily verified for the discrete case using a thought experiment. Let us say that we have a coin (not necessarily fair) and we want to assign probabilities to getting heads or tails. Any time we deviate from a 0.5-0.5 assignment we imply that one outcome is more likely than the other, indicating prior knowledge. If there is no prior knowledge, then there is no reason to assume that one outcome is more likely than the other. The 0.5-0.5 assignment assumes the least information over the fact that there are two and exactly two outcomes. There are situations where other statistics are known. For example, if only the mean of a distribution is known, we should select an exponential distribution. If mean and variance are known, then the normal distribution is the MED, and so on. Interested readers are referred to probability texts for further treatment of the topic.

6.3 Simulating Random Variables

Generating enough realizations of the input random variables is a major step in system simulation. How many realizations is enough depends on the problem. One way to check whether enough realizations have been acquired is to test the statistics of the random numbers generated. For example, when generating normally distributed random numbers, we would see if the mean and standard deviation of the generated numbers match those of the normal distribution we set out to generate. Furthermore, skewness and excess kurtosis should be close to zero. A computer code is usually employed to simulate random numbers. Most computer programs can generate uniformly distributed random variables with very precise statistics. There are many methods that can use these uniformly distributed numbers to generate realizations of other random variables if their pdf is known. Some of these methods are general, while others work only for certain distributions. In this section we present some of them.

6.3.1 CDF Inverse Method

The cumulative distribution function inverse method works by using a random number generator that can provide uniformly distributed random numbers between 0 and 1. For each of these numbers we find the inverse of the CDF of the random variable for which we want to generate numbers. The resulting numbers are distributed in the desired fashion. Let us assume that we want to generate Xs that are generated with the pdf given by $f_X(x)$. We first find the CDF of X as

$$F_X(x) = \int_{-\infty}^{x} f_X(t)\,dt \tag{6.6}$$

The CDF can also be acquired by fitting a closed-form expression if the above integral cannot be easily found. If the closed form of the CDF is known, we can skip this step. We then generate a suitable number of realizations, n, of uniformly distributed random variable Y, given by $Y_1, \ldots, Y_n$ such that $Y \sim U(0,1)$.

For each Y_i, we then find

$$X_i = F_X^{-1}(Y_i) \tag{6.7}$$

X_i's will be generated with the required pdf $f_X(x)$. As the number n increases, enough realizations will eventually be acquired to represent the distribution more and more accurately.

The intuitive explanation behind this method is as follows. Let us consider Figure 6.2, which shows a hypothetical CDF for a continuous random variable. As we can see, it is a single-valued, nondecreasing, continuous function. When we invert the function, for each point on the y-axis, realizations of a random variable on the x-axis are concentrated based on where the value of the pdf is the highest. Since pdf is the slope (derivative) of the CDF, it is evident that more points will proportionally be generated where the CDF has higher slope.

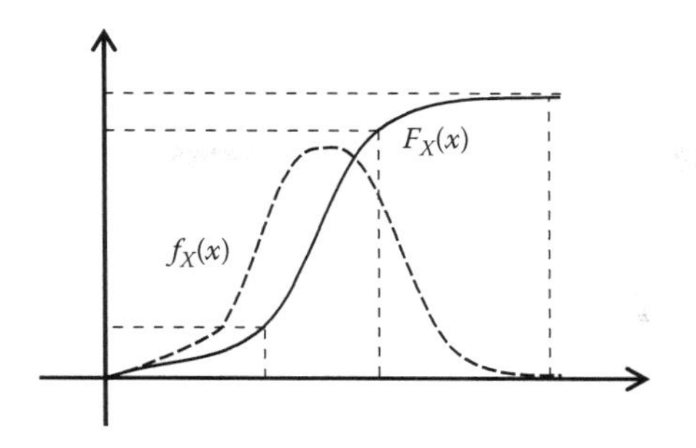

FIGURE 6.2
A schematic showing the intuitive explanation behind the CDF inverse method of generating random variables with a given distribution.

Example 6.2

Generate 20 random variables that are normally distributed with mean 2 and standard deviation 3.

SOLUTION

We first generate 20 uniformly generated random variables. Table 6.2 is an example. Each of the uniform random variables in the table is inverted using the standard normal CDF. Most statistical software, such as Excel and Minitab, find the inverse of the standard normal CDF. The resulting normally distributed variables need to be scaled to have the required statistics, as shown in Table 6.3.

We can verify the statistics of the numbers thus generated. The mean of the generated numbers is 2.06 and the standard deviation is 3.178. This error will reduce as we generate more realizations.

TABLE 6.2
Uniformly Distributed Seeds for Example 6.2

Realization	$Y \sim U(0,1)$	Realization	$Y \sim U(0,1)$
1	0.843180	11	0.246633
2	0.041460	12	0.349619
3	0.052548	13	0.927156
4	0.897148	14	0.968185
5	0.460336	15	0.742673
6	0.343415	16	0.397722
7	0.100318	17	0.554265
8	0.952282	18	0.347147
9	0.792720	19	0.201988
10	0.390938	20	0.319118

TABLE 6.3

Twenty Realizations of a Normally Distributed Variable with Mean 2 and Standard Deviation 3

Realization	$X = 3\Phi^{-1}(Y) + 2 \sim N(2, 3^2)$	Realization	$X = 3\Phi^{-1}(Y) + 2 \sim N(2, 3^2)$
1	5.022841	11	−0.055370
2	−3.201970	12	0.840952
3	−2.861900	13	6.364797
4	5.796401	14	7.564291
5	1.701238	15	3.954824
6	0.790518	16	1.222256
7	−1.83923	17	2.409333
8	7.002183	18	0.820896
9	4.447686	19	−0.503620
10	1.169375	20	0.589500

Example 6.3

Using the same seed as in Example 6.2, generate 20 random variables with a triangular distribution between 3 and 10, with most likely value at 6.

Solution

The CDF of the random variable is given by

$$F_X(x) = \begin{cases} 0 & x < 3 \\ \dfrac{(x-3)^2}{21} & 3 \le x \le 6 \\ 1 - \dfrac{(10-x)^2}{28} & 6 \le x \le 10 \\ 1 & x > 10 \end{cases}$$

Before we invert the CDF, we need to find the point where the function changes on the y-axis. This happens when $x = 6$ or $F_X(x) = 3/7 = 0.4286$. Therefore, while inverting the CDF, if the uniformly distributed seed is less than 0.4286, we will use the function $\frac{(x-3)^2}{21}$, while we will use the function $1 - \frac{(10-x)^2}{28}$ if the seed is greater than 0.4286. The corresponding values are shown in Table 6.4.

6.3.2 Box–Muller Method for Generating Normally Distributed Random Numbers

The Box–Muller method generates normally distributed random numbers using two uniformly distributed random numbers. The method is very useful when one does not have a way of inverting the normal CDF, for example,

TABLE 6.4

Twenty Realizations of the Triangular Distribution in Example 6.3

Realization	$X \sim T$ (3, 10, 6)	Realization	$X \sim T$ (3, 10, 6)
1	7.904538	11	5.275806
2	3.933092	12	5.709612
3	4.05048	13	8.571843
4	8.302986	14	9.056167
5	6.112766	15	7.315758
6	5.685464	16	5.890011
7	4.45144	17	6.467214
8	8.8441	18	5.700016
9	7.590884	19	5.05955
10	5.865257	20	5.588721

when one only has access to a scientific calculator. Even when using a scientific programming platform with access to the inverse CDF function, the Box–Muller method is preferred because it is much faster. Using two uniform seeds, u_1 and u_2, we get

$$z_1 = \sqrt{-2\ln u_1}\cos 2\pi u_2 \tag{6.8}$$

$$z_2 = \sqrt{-2\ln u_1}\sin 2\pi u_2 \tag{6.9}$$

Example 6.4

Using the first and second column seeds in Example 6.2 generate 10 realizations of two independent standard normal random variables.

Solution

Using the Equations 6.8 and 6.9 we get the variables shown in Table 6.5. Notice that the means and standard deviations of Z_1 and Z_2 will approach 0 and 1 for a larger sample. Also, the correlation coefficient will approach zero.

6.3.3 Generating Dependent Random Variables

Design variables affecting engineering systems are often correlated or, more correctly, dependent. This dependence has a significant impact on the performance of the system. For example, in electronic circuits, power spikes can cause the failure rates of downstream components to be correlated. In offshore structures, a hurricane subjects them not only to wind gust loads but also to storm surges.

TABLE 6.5

Standard Normal Variables Using the Box–Muller Method

Realization	Z_1	Z_2
1	0.012356	0.583949
2	−1.47815	2.044778
3	2.177519	−1.0726
4	0.456629	−0.09252
5	−0.05732	−1.24431
6	−1.17041	0.876221
7	−2.02104	−0.71709
8	−0.17924	0.256243
9	0.202511	0.650815
10	−0.57667	1.243327

Therefore, a design that treats these two forces as independent will underestimate the probability of failure. It is clear, therefore, that if we want to simulate engineering systems subjected to correlated random variables, we need ways to generate their realizations. In this section we present some methods that can be used to generate such random variables. We start with the most intuitive, the rejection sampling method, and then go on to more efficient methods.

6.3.3.1 Rejection Sampling

Rejection sampling is another method to generate random variables; its beauty is that it can be easily extended to jointly distributed random variables. The basic approach behind the method is that a large number of realizations of an easy distribution are generated, and then the realizations are accepted or rejected based on whether they correspond to the distribution whose realizations we require. It can be considered a brute-force method in that it does not require explicitly inverting the CDF; however, it does require a closed-form expression of the pdf or some way of calculating it. The appeal of the rejection sampling method is that any type of joint distribution with any correlation structure can be simulated. The downside is that it is not very efficient; it takes a long time to generate a handful of realizations. This inefficiency is exacerbated when a large number of jointly distributed variables are to be simulated.

The premise behind the rejection sampling method is that the likelihood of a variable to be in a region in the domain of the distribution (called support) is directly proportional to the value of the pdf in that region. An envelope distribution is chosen that is easier to generate, usually a uniform distribution (Figure 6.3). The envelope distribution should have a support (domain where the pdf is strictly positive) larger than that of the target distribution. Also, the value of the pdf should be greater than that of the target distribution. This, however, makes the area under the envelope distribution greater than 1. There is a simple work-around though, as we shall see.

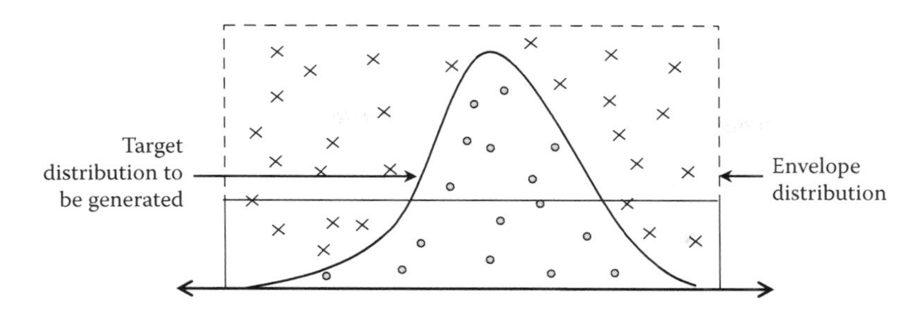

FIGURE 6.3
Rejection sampling method to generate random variables using an envelope distribution. Circles are accepted, while crosses are rejected.

The following steps are implemented to generate realizations of n jointly distributed random variables following an arbitrary joint pdf $f_{\mathbf{X}}(\mathbf{x})$:

1. Generate n random numbers $\mathbf{u} = \{u_1, u_2, ..., u_n\}$, each uniformly distributed between $u_i^{\min}$ and $u_i^{\max}$. The limits $u_i^{\min}$ and $u_i^{\max}$ are chosen so that they completely cover the support of $f_{\mathbf{X}}(\mathbf{x})$ within the box they define. The limits $u_i^{\min}$ and $u_i^{\max}$ can take different values for each dimension.
2. Evaluate the pdf of the joint distribution, $f_{\mathbf{X}}(\mathbf{x})$ at $\mathbf{u}$.
3. Generate another number u_t, uniformly between 0 and 1, and check if $Mu_t \leq f_{\mathbf{X}}(\mathbf{u})$, where M is equal to the maximum value that $f_{\mathbf{X}}(\mathbf{x})$ takes. This step ensures that the uniform distribution is properly scaled when comparing, but the area under the distribution stays 1.
4. If $Mu_t \leq f_{\mathbf{X}}(\mathbf{u})$, accept $\mathbf{u}$; that is, add it to the set of realizations generated so far. Otherwise, reject $\mathbf{u}$.
5. Repeat the four steps until enough realizations have been generated.

Rejection sampling works best with double-bounded distributions without sharp peaks. Sharp peaks increase the value of M, and as a result, a lot of uniform realizations are rejected in other regions. Clearly, one cannot generate (at least theoretically) unbounded distributions like the normal distribution. Approximations can be made by bounding the domain of the uniform distribution within, let us say, four standard deviations. Luckily, other methods exist to generate normal distributions. For other unbounded distributions, an approximate finite support can be chosen.

Example 6.5

Generate 20 realizations of the random variables given by the following joint pdf using the rejection sampling method:

$$f_{X,Y}(x,y) = \frac{x^3y^4(1-x)^3(1-y)^4}{0.00001134}, \quad X,Y \in [0,1]$$

TABLE 6.6

Twenty Realizations of the Random Variables for Example 6.5

Realization	X	Y	Realization	X	Y
1	0.448762	0.404631	11	0.22212	0.289152
2	0.413647	0.382352	12	0.498268	0.614382
3	0.714305	0.421032	13	0.764262	0.397314
4	0.749748	0.58697	14	0.415261	0.4917
5	0.642001	0.54277	15	0.562242	0.356799
6	0.687591	0.591451	16	0.509434	0.48029
7	0.770332	0.802714	17	0.422384	0.261713
8	0.211066	0.74293	18	0.551811	0.684626
9	0.263251	0.550697	19	0.528319	0.483928
10	0.339739	0.471826	20	0.377454	0.516459

SOLUTION

The joint pdf takes the maximum value when both X and Y are equal to 0.5, and the value is equal to 5.3833, that is, $M = 5.3833$. To generate realizations of the jointly distributed random variables, we start with generating uniformly distributed numbers on the support given by the box $[0,1]\times[0,1]$. For each tuple, we generate another uniform random variable, u, and test to see if $5.3833u$ is less than the value of the pdf at the tuple. If it is, we accept the tuple as a realization; otherwise, we reject it. This process was repeated until 20 realizations were accepted. Table 6.6 shows the results.

6.3.3.2 CDF Inverse Method for Jointly Distributed Variables

We discussed how rejection sampling can generate jointly distributed random variables relatively easily, but that it is inefficient when a large number of realizations are required. We can, however, extend the inverse CDF method to multivariate cases to make the process more efficient. Let us say that we need to generate realizations for variables in the vector $\mathbf{X}$ with a joint pdf given by $f_{\mathbf{X}}(\mathbf{x})$ and the CDF given by $F_{\mathbf{X}}(\mathbf{x})$. We will first need to find the marginal CDF of one of the variables in the vector $\mathbf{X}$. For a two-variable case where $\mathbf{X} = \{X_1, X_2\}^T$, the marginal CDF of X_1 is given by

$$F_{X_1}(x_1) = \lim_{x_2 \to \infty} F_{\mathbf{X}}(x_1, x_2) \tag{6.10}$$

Similarly, the marginal CDF of X_2 is given by

$$F_{X_2}(x_2) = \lim_{x_1 \to \infty} F_{\mathbf{X}}(x_1, x_2) \tag{6.11}$$

The simulation method works as follows. Using the inverse CDF method and the marginal distribution of X_1, we generate a realization of X_1 as

$$x_1^i = F_{X_1}^{-1}(u_i) \tag{6.12}$$

Now we find the conditional distribution of X_2, given that the value of X_1 is x_1^i. This is given by

$$F_{X_2}\left(x_2|X_1 = x_1^i\right) = \frac{F_{\mathbf{X}}(\mathbf{x})}{F_{X_1}(x_1)} \tag{6.13}$$

Using the conditional CDF of X_2, we can generate a realization x_2^i , and as a result, we have one pair of realizations, $\{x_1^i, x_2^i\}$. The process is repeated until enough realization tuples have been generated. Notice that this method can be extended to many variables, and for each successive variable we will find conditional distribution on the variables whose realizations have already been found.

6.3.3.3 Generating Dependent Normally Distributed Random Variables

If we have access to independent normally distributed random variables (we have already presented a few ways to do this), we can perform a simple transformation to generate correlated normally distributed random variables. Assume that we need two normally distributed random variables with correlation coefficient* ρ. Starting with two independent normal variables X_1 and X_2, define X_3 as

$$X_3 = \rho X_1 + \sqrt{1-\rho^2}\,X_2 \tag{6.14}$$

Now X_3 is normally distributed and has a correlation coefficient of ρ with X_1.

6.4 Copulas

Copulas are a very powerful way to represent the dependence between random variables. Recall that a joint cumulative distribution for a set of random variables, $\mathbf{X} = \{x_1, x_2, \ldots, x_n\}$, is defined as

$$F_{\mathbf{X}}(x_1, x_2, \ldots, x_n) = P(X_i \le x_i; i = 1, \ldots n) \tag{6.15}$$

* Notice that for a pair of normally distributed variables, the correlation coefficient captures all the information about the dependence between them.

The individual random variables x_i's themselves follow distributions represented by the cumulative density functions $F_{X_i}(x_i)$. As we discussed earlier, these single-variable distributions are referred to as marginal distributions or *marginals*. For a function to qualify as a joint distribution function, it has to satisfy the following four conditions:

1. $\lim_{x_i \to -\infty} F_X(x_1, x_2, \ldots, x_n) = 0 \quad \text{for any} \quad i \in \{1, \ldots, n\}$
2. $\lim_{x_i \to \infty} F_X(x_1, x_2, \ldots, x_n) = 1 \qquad \forall i \in \{1, \ldots, n\}$
3. $\lim_{x_i \to \infty} F_X(x_1, \ldots, x_j, \ldots, x_n) = F(x_j) \quad \forall i \neq j$
4. $F_X(\cdot)$ is n-increasing

The first three conditions are pretty straightforward; condition 4, however, needs a little explanation. An n-increasing function is the n-dimensional version of an increasing function in one dimension. In two dimensions, a 2-increasing function, F, is such that

$$F(a_2, b_2) - F(a_1, b_2) - F(a_2, b_1) + F(a_1, b_1) \geq 0 \quad \forall a_2 \geq a_1, b_2 \geq b_1 \tag{6.16}$$

Realize that the n-increasing property is *not the same* as being increasing in all n dimensions individually. For example, the function $\max(x_1, x_2)$ is increasing in x_1 and x_2 but is not 2-increasing. Similarly, the function x_1x_2 is 2-increasing but is not increasing in x_1 (or x_2) for negative values of x_2 (or x_1). The n-increasing function property guarantees that for any *box* in n dimensions, the probability that the random variable lies within it is greater than or equal to zero (probability cannot be negative).

A copula, usually denoted with letter C, is simply a joint distribution that maps $[0,1]^n$ to $[0,1]$ and satisfies the following conditions:

1. $C(0, \ldots, 0) = 0$
2. $C(1, \ldots, 1) = 1$
3. $C(1, \ldots, 1, u_m, 1, \ldots, 1) = u_m \quad \forall m \leq n$
4. C is n-increasing

Properties 1, 2, and 4 follow directly from our discussion above. Property 3 simply says that the marginal distributions are uniform.

6.4.1 Sklar's Theorem

Sklar's theorem (1959) is an important result in probability theory. Sklar's theorem highlights the central role copulas play in combining information from marginal distributions into a multivariate joint distribution. The theorem states that any joint distribution $F_X(x_1, x_2, ..., x_n)$ can be represented using the marginals and a copula; in other words, for a two-variable case,

$$F(x_1, x_2) = C(F_1(x_1), F_2(x_2)) \tag{6.17}$$

A copula therefore models the dependence information between the random variables independently from distributions of the individual variables themselves.

6.4.2 Constructing Copulas

Nelsen (2006) provides a detailed description of how copulas can be constructed. Using copula generator functions is one of the most common ways of constructing copulas. For a two-variable case, a strict Archimedean copula is defined as

$$C(u, v) = \varphi^{-1}(\varphi(u) + \varphi(v)) \tag{6.18}$$

where φ is a continuous, strictly decreasing function from $[0,1]^n$ to $[0,\infty]$ such that $\varphi(0) = \infty$ and $\varphi(1) = 0$. The function φ is called the generator of the copula. A famous one-parameter Archimedean copula is Frank's copula, defined as

$$C(u, v) = -\frac{1}{\theta}\ln\left(1 + \frac{(e^{-\theta u} - 1)(e^{-\theta v} - 1)}{e^{-\theta} - 1}\right), \quad \theta \neq 0 \tag{6.19}$$

and the generator of Frank's copula is

$$\varphi(t) = -\ln\left(\frac{e^{-\theta t} - 1}{e^{-\theta} - 1}\right), \quad \theta \neq 0 \tag{6.20}$$

One of the advantages of Frank's copula is that it can model extreme positive as well as extreme negative dependence between the random variables. In many problems, marginal distributions are known with much better accuracy than joint probability distribution because a small number of assessments is required to assess distributions over single variables. As the number of variables increases, the number of assessments required increases exponentially. Copulas can provide a simple way of combining the information from these distributions into a joint distribution when the dependence information is acquired separately, and also allow us to investigate the sensitivity to the dependence information. The parameter θ in Frank's copula determines the degree and nature of dependence between the variables. A positive value of θ signifies a positive dependence, while a negative value signifies negative dependence. A value of 0 means that the variables are independent.

6.5 Note on Statistical Dependence and Independence Conditions for Multiattribute Utility Function Assessments

Recall that in Chapter 3, we discussed preferential and utility independence conditions. These conditions, when satisfied, allow us to combine single-attribute utility functions into a multiattribute function using a multiplicative form. We take this opportunity again to remind the reader that the statistical dependence between two attributes has nothing to do with the preferential and utility independence conditions. Two attributes could be statistically dependent and still be independent in the decision maker's mind, and vice versa. In fact, attributes are generally expected to be dependent; otherwise, one would simply improve one of the attributes without affecting the others.

Example 6.6

Using simulation, show that the decision maker with the following utility function is risk-averse:

$$U(x,y) = \sqrt{xy} \quad x,y \in [0,10]$$

TABLE 6.7

Twenty Realizations of the Random Variables for Example 6.6.

	x	y	$U(x, y)$
	5.65367	2.210755	3.535375
	7.659998	2.789646	4.622627
	8.70043	6.039471	7.248862
	0.812313	3.628162	1.716742
	4.102897	0.619502	1.594287
	9.09058	5.355821	6.977645
	4.104981	8.30311	5.83816
	1.562921	2.72511	2.063767
	6.817634	7.800699	7.29262
	7.655582	8.955475	8.280058
Mean	**5.616101**	**4.842775**	**4.917014**
Function value at the mean of arguments			**5.215123**

SOLUTION

We generate a few points in the domain of the two variables (does not matter what distribution) and check to see if the expectation of the function is lower than the function value at the expectation of the variables.

Table 6.7 shows that it is indeed the case; that is, the utility function is concave, pointing to risk-averseness. While this result shows that the utility function is generally risk-averse, no guarantees of global convexity or concavity can be made. A simulation-based test like this one should be resorted to when there is no other way of verifying concavity, for example, the Hessian test for negative semidefiniteness.

Problems and Exercises

1. Why do we simulate engineering systems?
2. Describe the Monte Carlo method. What is its main advantage? What is its main disadvantage?
3. Explain the intuitive premise behind the inverse CDF method of generating random variables. Can you also prove mathematically that the method works?
4. Generate 20 realizations of each of the following distributions using the inverse CDF method:
   a. Normal distribution $N \sim (1, 3^2)$
   b. Triangular distribution with lower and upper limits at 2 and 10, with the most likely value at 4
   c. A variable that is drawn 50% of the time from a normal distribution with mean 0 and standard deviation 1, and the rest of the time from a uniform distribution between –1 and 1
5. Find the best-fit beta distribution based on the following realizations:

0.025722	0.402802	0.453853	0.061067	0.34114
0.003551	0.875539	0.642026	0.925341	0.011989
0.612129	0.462609	0.11858	0.06771	0.060763
0.32099	0.888867	0.619814	0.000879	0.205898
0.009844	0.078703	0.863516	0.573151	0.614894

6. Generate 20 realizations of a standard normal random variable using the Box–Muller method.
7. Discuss the Bayesian method for uncertainty modeling.

8. What is a conjugate prior?
9. What is the maximum entropy method? What is the main reason behind using the maximum entropy method when limited information about a distribution is available?
10. Describe the premise behind the rejection sampling method. When is it most commonly used?
11. How would you generate correlated normal random variables?
12. Generate 200 realizations each of two correlated standard normal variables (mean 0 and standard deviation 1) when their correlation coefficient is 0.6.
13. Outline the steps to generate dependent random variables when their joint CDF is available.
14. What is a copula? Why are copulas important?
15. What is Sklar's theorem? Discuss how it is relevant when dealing with copulas.
16. Determine if the decision makers with the following utility functions are risk-seeking, risk-neutral, or risk-averse, using simulation:

    a. $U(x,y) = x\sqrt{y} \quad x,y \in [0,10]$

    b. $U(x,y) = x^2 + y^2 \quad x,y \in [0,10]$

    c. $U(x,y,z) = \sin x \cdot \sin y \cdot \sin z \quad x,y,z \in [0,\pi/4]$

7

Product Development and Systems Engineering

7.1 What Is Product Development? What Is Systems Engineering?

Product development (PD) is the sum total of all activities that are undertaken to identify the need for a product, determination of its functional and aesthetic shape, successful production, and commercialization. Product development has many facets that need to be understood by an engineer before the technical engineering design can begin. Formal understanding of the entire PD process is therefore very important. In Chapter 2, we established that the arrow of causality flows not from technical engineering design, but to it from the financial goals of the company. Engineering design, fortunately or unfortunately, is almost never an end in itself. It is done to achieve a larger goal, which most of the time is maximization of profit for a firm in the short or long term by developing profitable products. Consequently, engineering design knowledge coupled with understanding of business aspects makes for a very successful engineer. An engineer who understands product development is more successful as part of a company or even as an entrepreneur.

In this chapter we will cover the major concepts encompassing product development. We will do so in the context of engineering design. We will learn how engineering, while not the causal force, does inform each and every aspect of product design and is key to differentiation in products and competitive advantage. For most engineers, finding the core element of engineering design in the PD process may initially seem difficult. They may even be initially overwhelmed with due diligence and nontechnical issues such as sales, demand modeling, customer service work, and other issues that seem at best extraneous to them. This is because they are traditionally trained to have a cookie-cutter approach to engineering design: find the best possible engineering solution to a problem. They should understand that knowledge of product development concepts only enhances their technical knowledge. Furthermore, as we shall see, product development has drawn heavily from highly technical results in mathematics and marketing literature, and an engineer does end up using all knowledge acquired as part of his or her technical training.

Systems engineering is an allied science to product development. The International Council on Systems Engineering (INCOSE) defines systems engineering: "Systems Engineering is an interdisciplinary approach and means to enable the realization of successful systems. It focuses on defining customer needs and required functionality early in the development cycle, documenting requirements, then proceeding with design synthesis and system validation while considering the complete problem." As such, boundaries between systems engineering and product development appear fuzzy. We consider systems engineering as a means to enabling a successful product development process. When implemented properly, systems engineering becomes a major subset of product development in that it provides tools that prove critical in designing parts that form the system, as well as its interactions with the environment. Furthermore, it focuses closely on the engineering aspects of systems design, formalizing and integrating the whole process with the profit-making endeavors of the firm in question.

The goals of a PD process are many: to meet a market need, to establish or enhance competitive advantage, to streamline design and manufacturing processes, to adhere to legislative regulations, or any combination of these. A company has a successful PD process if there is continuous improvement in the products and services offered to the customers, sufficient adaptability to the needs of the customer, the wherewithal to implement a new product design or concept, and the ability to provide products at cheaper prices and in a timely fashion—all the while maintaining integrity within the teams in the organization and showing environmental and societal stewardship.

7.1.1 Industrial Design

In many ways, the topics in this chapter will be similar to the design methodology we discussed in Chapter 2. In this chapter, however, we will take a slight departure from the classical notion of engineering design and include tenets of *industrial design*. Industrial design is rooted in the *profitable* manufacture of products that *customers want*. As a result, aesthetics, ergonomics, and user-friendliness also become attributes of concern. As an example, a particle accelerator does not need industrial design as much as a vacuum cleaner does. Indeed, concepts in this chapter can be readily applicable to the design of a particle accelerator; the challenges faced are totally different and the monetary gains are almost never the motive. Money (or more succinctly, profit), on the other hand, is the driver of industrial design—hence the focus on success in the marketplace. There is nothing inherently evil in this approach because profitable companies contribute to society by providing products and services as well as employment.

Industrial design involves making a product that goes beyond meeting the functional requirements. Industrial design actively considers the human user's interaction with the product. It defines a form that facilitates the human user's use of the product. It also takes into account ergonomics so that the use of the product is not cumbersome and does not lead to user fatigue. The product must also be safe to use, and if not, must carry proper warnings outlining conditions where it becomes unsafe. If commoditized, the product should also have aesthetic appeal, which might prove to be an advantage over the competitors'. The functioning of the product should be intuitive so that the end user can start using the product, or at least its basic functionalities, with minimal training. Industrial design also helps create a brand image that goes beyond profitability. It establishes the product in the customer's mind. For example, one can recognize a general theme in the different car variants made by a particular manufacturer. Similarly, many electronics companies tend to spend a lot of effort in making their product immediately recognizable. The consumer electronics company Apple, for example, has successfully created a brand image based on easily recognizable products and careful industrial design.

Last but not least, industrial design also informs engineering design. In consumer products, unless an entirely new product with entirely new technology is offered, customers have some expectations on what the product's form is. Therefore, form introduces design constraints that must be sorted out. An example is the battery in a cell phone. Customers expect a certain battery life from a cell phone in addition to the form requirements that the phone not be bulky. Miniaturization of batteries is a big technical challenge that must be overcome. Therefore, in this case industrial design informs engineering design. One must understand that even if the customer does not have form/user interface expectations, a product development process should still consider them.

7.2 More Roadblocks to Engineers Thinking about Product Development

Engineers in general tend not to think much about product development and mostly focus on their engineering domain. Chapter 1 identified some roadblocks. Here we elaborate on some of them in the context of product development and also discuss a few additional ones.

1. **The engineer does not know everything he or she needs to know:** There are two aspects of lack of knowledge in engineers about product development. First, the traditional technical courses in engineering rarely cover in detail the seemingly extraneous factors that are almost always part of practical product development. Second, in a corporate setting, engineers are often segregated from the product

development process. As a result, engineers rarely understand why they are doing what they are doing. Furthermore, within the technical team, the output of their efforts is measured in technical accomplishments as opposed to an increase in profitability or similar metrics.

2. **The engineer is not interested in product development and only enjoys his or her domain:** Many times engineers become engineers because of their love for the technical knowledge. Engineers then tend to make design decisions based on what is technically novel rather than what the product needs given the cost and functionality constraints.
3. **Lack of confidence of the management in the engineer's skills:** Sometimes engineers are not entrusted with product development roles because higher management does not trust the engineer's skills. Ironically, most successful managers started out as engineers and had to deal with the same attitude from their superiors. The educational and business environment has fortunately changed somewhat in that many engineering schools now offer business-oriented courses to engineering students.
4. **Organizational problems:** Larger companies sometimes have clearly defined departments and technical teams that are physically separated (even geographically separated over huge distances). An example would be the large commercial airplane manufacturers that design and manufacture aircraft subsystems at various locations. Of course, the aircraft subsystems should work together, when they are assembled together. An error in coordination of various facilities or mistakes in implementations can have far-reaching consequences in success of design projects. Despite this, in such organizations engineers are more likely to focus on their domain versus the overall product.
5. **Ill-defined problem statement:** A problem can be ill-defined. While there can be no general recipe for identifying what makes an engineering problem well defined, there are signs to be seen. If design specifications violate obvious engineering constraints, require major changes to an existing design, ask for vague attributes, or do not specify critical information, the engineering decision makers should backtrack and define the problem more clearly. The more questions they ask, the better they will be able to place their efforts in the context of the product design.
6. **Team thinking:** Advanced degrees are usually designed to promote independent thinking. Engineers take this mindset to a company and do not necessarily share ideas or seek advice. Successful product development is all about team thinking.

7.3 Steps in Product Development

In this section we go over a generic product development process that is applicable to many different product types. Companies differ in their approach to product development. There are companies that have a "flat" structure in that almost all decision makers are approachable. They get together in a kickoff meeting and brainstorm the development of a new product. Technical, form, and function deliverables are discussed along with the hard constraints, such as money allocations and development time. The end result of the meeting is a (hopefully) clear understanding on what needs to be done by everyone involved. Such an amorphous method neither works in a large organization nor gives us much to talk about in terms of a systematic study of product development. Large organizations approach the product development process in a very systematic way and have specific teams focused on each subissue—be it technical, marketing, sales, software, or organizational. Therefore, we will approach product development as it would take place in a large, differentiated organization. It also makes sense to learn product development concepts in a domain-independent fashion, that is, understand the general approach that a large engineering firm should use mostly independently of the technology and product in question. Ulrich and Eppinger (1999) define six phases of a generic product development process, and we follow their description here:

1. Planning phase
2. Concept development phase
3. System-level design phase
4. Detailed design phase
5. Testing phase
6. Production ramp-up

They define the planning phase as the one where corporate-level decisions are made regarding funding allocations and market studies for new product development. A new product may be the result of a perceived void in the marketplace or due to a clear order from a customer. A clear order takes away a lot of uncertainty from the product development process in terms of product attributes, cost, and shipping dates. Even with the lack of a clear order, a company must also continually look to improve its existing products or offer new ones to maintain market share and gain competitive advantage. Voids in the marketplace can be identified through surveys, which we will discuss later in the chapter. In the planning phase, top decision makers also allocate resources, both manpower and monetary, to the project. The planning phase also includes understanding customer needs. To this effect, the results from the surveys or the customer needs from the product (in case of a clear order) need to be interpreted and put in a form in which they can be addressed in a systematic way. Quality function deployment is a way to encode customer needs and translate them into functional requirements. We will discuss this in detail in the next section.

In the concept development phase various product design alternatives are proposed and considered. Recall from Chapter 2 that product design is often decomposed into design subproblems (DSPs) so that they can be individually tackled. This is generally done in a group setting. Brainstorming is an example of a method used in the concept generation and development phase. Technical challenges posed by the design alternatives are considered, and solutions proposed. Team members also consider how the solutions to DSPs come together in a product. While most ideas are welcome during the concept generation phase, by the end of it, technically infeasible ones are rejected. The basics of the product start taking shape, and therefore documentation becomes critical. It is important to have rough drawings and specifications of all the rough designs proposed.

The physical design of the product is considered in depth, in the system-level design phase. All the sources of energy, mechanisms of operation, user interface, and expected cost are also decided, along with tolerances and uncertainties. In the detailed design phase, enough functional and geometric information is available that the product prototype can be built. The testing phase involves production of some prototypes that are tested for functionality and their ability to satisfy functional requirements. The product development process iterates whenever there is a need to backtrack and improve the design. After testing, and any subsequent improvements, the product is put into mass production. The initial production run is usually slow because it involves training of manpower and general streamlining of production methods. Once the kinks in the process are sorted out, the mass-produced product is available for distribution to end users.

For the purpose of this chapter, we will touch upon these steps independently and not necessarily in the order mentioned. This is done to help the reader grasp these concepts independently and put them in the context of decision based design because additional relevant topics are also presented. The central idea in this chapter is that even though product development is a collection of disparate ideas, it has a central goal: to profitably realize products by making good decisions.

Before we proceed, let us go over two definitions:

1. **Customer needs:** Customer needs refers to an almost verbatim requirement from the product as requested by the customer. Customer needs are an essential input to the PD process. An example would be "The product should be able to clean a carpet of accumulated dust particles."

2. **Functional requirements:** Functional requirements refer to the attributes of a product that enable it to address customer needs. All customer needs must be addressed by at least one functional requirement. An example would be "The product generates enough suction to remove accumulated dust particles from the carpet"—as in a vacuum cleaner.

7.3.1 Customer Needs Assessment

A company survives by making more money than it spends. A company that sells products or services needs to make sure that what it offers is valuable to the customer who buys them. Very few companies can be (or have been) successful trying to produce something and then worrying about how to get customers to buy it. The process of realizing what the customer wants is called customer needs assessment. In this section we discuss the philosophy behind accurately capturing what the customer wants, and then we will discuss some practical methods to go about doing it.

7.3.1.1 Survey and Data Collection

Products are designed to be successful in the marketplace so they can generate revenue for a company. For products that have been around for some time, there is generally information on customer preferences that the company has acquired in the past. For new products or before making major changes to a product, these preferences must be acquired anew. It must be assumed that customers buy only what is valuable to them, and surveys are a way of finding out what the customers value in a product. A survey is performed on a subset of the target customers for the product. Surveys can be performed as conducting in-person interviews, observing purchasing behavior, asking questions over the Internet or telephone, or conducting focus group studies.

Surveys are often incentivized with gifts or monetary compensation to ensure customer participation. However, they do not guarantee accurate responses. In-person interviews might include approaching a customer in a store or making an appointment over the phone to meet at a prespecified place and time. Many stores run loyalty programs (e.g., magnetic swipe cards) where customers get discounts if they agree to sharing personal information. It helps the stores maintain huge repositories of information through these loyalty programs. Many times surveys are conducted over the phone or over the Internet, and usually reward in the form of discounts or access to otherwise paid sections of websites. The most demanding way of surveying customers, in terms of both expense and time, involves focus groups. In a focus group a carefully chosen section of customers, usually matching the demographic of the target customer group, is selected. The group meets for a few hours, for which it is compensated, and provides answers to targeted questions about the product proposed. There are many benefits of a focus group over other methods of customer data collection:

1. Since customers are compensated, they feel as if they have a duty to provide honest answers.
2. Customers typically choose to come to these meetings and, therefore, are less likely to get distracted by phone calls or other commitments.
3. Customers can interact with each other. This helps them build upon each other's responses.
4. It is unlikely that a relevant customer need will be missed if the session is well conducted for a long enough period of time.

It is generally seen that a focus group of 20 to 30 customers is enough to fully understand customer needs from a product. Many times companies hire a third-party firm to perform focus group surveys for them. This not only helps in maintaining anonymity, but the third party is also typically more experienced in performing surveys.

Well-designed surveys help customers think critically about a product's attributes and provide answers that best reflect their preferences. Questions in a well-designed survey also do not overwhelm the respondents because then customers are more likely to give incorrect answers. One should avoid questions where the answer is already known. For example, if customers are asked questions about an automobile, questions such as "Do you like a safe car?" do not add anything to our knowledge because the prevailing answer will be yes. At the same time, a question such as "Would you purchase a car that gets five-star ratings in crash tests but has a slightly less fuel economy and costs $2,000 more than your current car?" involves three attributes, which may overwhelm them. This is not to say that their trade-off behavior over multiple attributes should not be assessed. Every question should be well thought out and should add to the company's knowledge of the customers' needs. As an example, once the customers have provided answers on safety requirements, fuel economy, and cost, and are comfortable thinking about the three attributes, they will be much more comfortable answering the second question.

Surveys should not be just about bombarding the customers with questions. It is equally important to listen to their issues with the current version of the product or what competitors have to offer. In a focus group, letting customers discuss their expectations from a new product out loud can be very informative. Very often it is seen that customers not only mention what they need, but have also thought about how to design a product that will satisfy their needs. Needless to say, this makes the job of the manufacturer that much easier. Many times a new dimension to product functionality is revealed based on innocuous discussion among customers. Such discussion therefore must be encouraged. Ulrich and Eppinger (1999) also advocate using props and pictures to demonstrate the product concept and other competing products that are out there. These will clearly help customers better understand what the product does and how, particularly what its physical dimensions are. Many times ergonomic issues with the product when an inexperienced customer tries to use it for the first time are identified.

Good documentation of raw data acquired from the customers should be maintained for future reference. All responses from the customers, including sketches and even body language while using the product prototypes, should be noted and stored. One could also video-record the whole focus group session, if possible. Customers do not always provide information in a form that is directly usable in product design. The raw data, once collected, need to be modified to be usable. Many times customers provide the same needs in many different ways; an engineer should compare and eliminate duplication. The engineer should also not get fixated on a design implementation but rather focus on the functional requirement from the product. Finally, customer statements should be translated into customer needs, with careful attention to their relative importance.

Surveys should be conducted on customers most representative of the product's target population. For most products, lead users, or the customers who are the first to try a new product, can be identified. They are generally more willing to take part in focus groups and other marketing surveys. Lead users in a technology area such as electronics are at the forefront of technology and are very enthusiastic about new products. For other products, lead users could be individuals who use the product regularly and are dissatisfied with it. In both cases, they are valuable resources because they have already identified what is needed from a new product. Another aspect sometimes overlooked in surveys is whether the customer is in a position to make the buying decision. This happens routinely in the case of products targeted toward children, and even in the case of companies that buy products that their employees use, for example, work computers. Surveys in such cases should try to understand what it would take to influence the actual decision makers.

As a note on leasing, many times manufacturers can and do enter into a leasing agreement with the customers who agree to pay a certain amount of money for a service; that is, they will always have access to the product that the manufacturer will maintain/replace for a certain fixed amount. In such cases, preferences of the particular customer become paramount for a successful leasing agreement. Formal decision analytic methods should be used in these cases. We discussed utility function assessment methods in Chapter 3. Surveys on focus groups should also explore these alternative product-selling ideas.

7.3.1.2 Conjoint Analysis

Conjoint analysis is a way of determining which combination of attributes is most attractive to customers, thus predicting their decision-making behavior. This method works by determining "part-worths," or the value of a marginal

TABLE 7.1

Attribute Levels for the CFL Lamp

	Light Output	Wattage	Price
Low	500 lumens	8 W	$3.00
Mid	700 lumens	10 W	$4.00
High	900 lumens	12 W	$5.00

change in one attribute of a product. This can be compared with the value in marginal change in another attribute or attributes if their part-worths are also known. We describe this method using an example. Consider a manufacturer that is trying to market a new compact fluorescent lamp. The bulb has three attributes: light output (L), wattage (W), and price (P). The preference order is such that customers prefer higher light output to lower, lower wattage to higher, and lower price to higher. The attribute levels are defined in Table 7.1.

Of course, a customer will like a lamp with 900 lumens output and 8 W power consumption at $3.00 price. However, high-output, high-efficiency lamps are likely expensive to manufacture, and the manufacturer might not be able to sell them for $3.00. A manufacturer therefore would like to know the ideal attribute levels that are cost-effective to produce and that the customer is likely to buy. The customer, in this example, will be asked to rank different CFL bulbs with different attribute levels. Based on the responses, numerical weights can be fitted to each attribute so that the same rankings are reproduced by a value function. Assume that the customer gives the rankings shown in Table 7.2 for the attributes of light output, wattage, and price.

Notice that not all combinations of the attribute levels are used. Given the rankings provided by the customer, the next step in conjoint analysis is to find the part-worths associated with each attribute, which determine their

TABLE 7.2

Rankings Provided by the Customer for the CFL Lamps

Rank	Light Output	Wattage	Price
1	900 lumens	8 W	$5.00
2	900 lumens	10 W	$5.00
3	700 lumens	8 W	$3.00
4	900 lumens	12 W	$3.00
5	900 lumens	12 W	$5.00
6	700 lumens	10 W	$4.00
7	500 lumens	8 W	$3.00
8	700 lumens	12 W	$4.00
9	500 lumens	12 W	$3.00
10	500 lumens	12 W	$5.00

TABLE 7.3

Part-Worths Associated with Different Attributes for the CFL Lamp

	Light Output		Wattage		Price	
	Value	**Part-Worths**	**Value**	**Part-Worths**	**Value**	**Part-Worths**
Low	500 lumens	0	8 W	10	$3.00	4
Mid	700 lumens	10	10 W	5	$4.00	2
High	900 lumens	20	12 W	0	$5.00	0

relative importance in the minds of the customer. There is some arbitrariness involved in this step. Assuming that the overall value from the CFL lamp to the customer is a linear sum of the part-worths of the attributes, one scheme that works is given in Table 7.3.

Realize that the rankings in Table 7.2 can be achieved using many different combinations of associated part-worths, leading to some arbitrariness. There are many further techniques in conjoint analysis that can deal with this issue, but we will not study them in this book. The simple method shown can be a good first-cut method to compare two options. Therefore, if the manufacturer were to compare two bulbs with attribute levels at {700 lumens, 8 W, $4.00} and {500 lumens, 10 W, $3.00}, we would first find the values associated with the two bulbs. Using the part-worths in Table 7.3, the first bulb is valued at 10 + 10 + 2 = 22 units, and the second bulb is valued at 0 + 5 + 4 = 9 units. Therefore, the customer would value the first bulb more. If the two bulbs have the same profit margin, the first one should be manufactured.

7.3.1.3 Compatibility with Decision Analysis

Conjoint analysis has some elements of decision analysis. A value function is usually acquired (similarly to the linear sum above) from the conjoint analysis that measures the overall worth of a product by modeling the trade-off between the attributes. In decision analysis this role is performed by multi-attribute utility functions. The difference is that instead of assessing utility functions directly, the customer provides rankings, which people are known to be more comfortable providing. The two major shortcomings of the value function in conjoint analysis are

1. Uncertainty cannot directly be taken into account, something that utility functions by definition do. In the above example, if the attribute levels were uncertain, one could not make a decision just by using the part-worths.
2. The strengths of preferences calculated in the part-worths almost always have some arbitrariness in them. While advanced methods have been developed to get around this issue, they still do not match the rigor of decision analytic methods.

Conjoint analysis is still used widely because of its simplicity. When limited data are available, precluding a well-modeled utility function, conjoint analysis can be used to make rough assessments.

7.3.1.4 Stated Preferences vs. Revealed Preferences

Two potential pitfalls of surveys, or for that matter, any method to collect customer data, are related to correctness and scalability. Correctness, in terms of preference assessment, refers to how accurately a survey captures the preferences of the customers. Scalability refers to how easy or *defensible* it is to extrapolate the preferences of the customer to situations that were not explicitly mentioned in the survey questions.

Correctly modeling preferences involves extracting as much information from survey responses as possible while minimizing errors and extrapolations. Clearly, better-phrased questions lead to more informative responses. A question phrased as "Do you like environmentally benign cars?" will get mostly yes answers, but it is something we knew already. A more relevant and useful question would be "Will you purchase a car that has a fuel economy of 30 mpg and costs $25,000?" The second question not only puts numbers to vague quantities, but also makes the customer actually think about the purchasing decision.

Scalability must be carefully considered when extrapolating information acquired from surveys. Some techniques such as conjoint analysis appear easily scalable, but it is unclear how defensible such scaling would be. In the light bulb example, can we extend the part-worths values to light output levels beyond 900 lumens? Theoretically yes, we can fit a polynomial function to the part-worths as a function of the light output and extrapolate. Such scaling will almost always lead to erroneous answers, particularly when other attributes also go beyond the levels presented to the customer. Easy scalability does not imply defensible scalability. Whenever an inference is to be made that goes beyond what was explicitly stated in a survey, extreme care should be taken.

This brings us to concepts of stated and revealed preferences commonly mentioned in the research literature. Stated preferences refer to the preferences of customers acquired by directly asking questions about the attributes they desire in the product. Revealed preferences are acquired by observing and understanding the decisions that customers make in real life. Preferences that are collected through revealed preferences are notoriously hard to extrapolate. Acquiring stated preferences in a decision based design context would involve asking what attributes a decision maker cares about, what their ranges of negotiability are, and finally, the lottery questions to assess the utility function, which includes their attitude toward risk. To model revealed preferences, customer purchasing behavior is observed (an example is the frequency cards given to customers in coffee shops) or customers are directly asked about their past purchases. A utility functional form is assumed, and the parameters are fit using regression or any other technique. We covered these in detail in Chapter 3 on decision analysis. A good survey technique requires understanding of the limitations of the stated and revealed preferences and asking questions that will establish a value or utility function within the whole decision-making frame.

7.3.1.5 Quality Function Deployment

Here we briefly touch upon quality function deployment (QFD), a method used to translate customer needs into functional requirements in the product. A common tool in quality function deployment is the house of quality (HOQ). It is a technique to formally assess, encode, and represent the preferences of the customers in an easily readable format. Figure 7.1 shows the schematic of an HOQ. An HOQ acts as a written-down repository of the design steps taken by a design team. Pictorially, it resembles a house, and hence the name. It includes a list of customer needs (CN) statements along with their relative importance as indicated by the customer. As shown in the figure, this information forms the left-hand part of the "house." The product design attributes (functional requirements, FR) that will help meet them are written alongside these needs (the middle box). Each customer need must be met by at least one functional requirement. An important consideration in fulfilling a functional requirement is understanding how it was done in the past or how competitors do it now. Information on this is written down in the box on the right. Therefore, following a row from left to right, one would see an objective requirement from the product, how it can be addressed, and

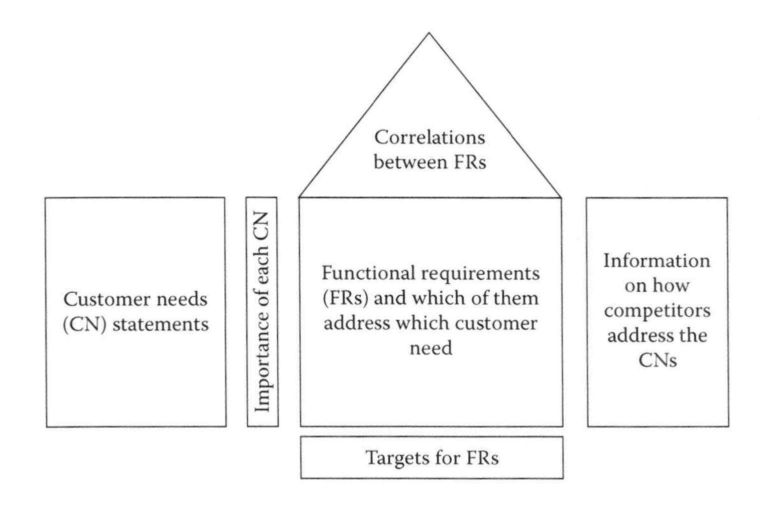

FIGURE 7.1
A schematic of a house of quality, a common way to represent customer needs and their fulfillment through functional requirements.

how it has traditionally been addressed (or by competitors). An HOQ also stores information on how different functional requirements are correlated, in the triangular box on the top of functional requirements in what forms the roof of the house. Many further enhancements exist, which help encode more information than the simplistic figure shown, but we refer the reader to dedicated books on the topic.

7.3.2 Creativity in Engineering Design

Creativity plays a critical role in engineering design, especially new product design. It can also just as easily be a part of incremental improvement in a product. Creativity is an abstract concept as it relates to the ability to provide original, useful, and substantially new ideas or solutions. It has the potential to cause fundamental improvement in a product, leading to competitive advantage in the marketplace. Contrary to popular conception, creativity is not all innate and can be learned. In this book we create a distinction between creativity from a novice and that from an experienced engineer. To a fresh engineer, product development seems a very haphazard process with no clear direction. He or she may feel intimidated by the number of (successful) creative solutions proposed by his or her seniors who are more experienced. The fresh engineer should realize that he or she can be creative as well, but his or her approach to creativity is and should be different from that of an experienced engineer.

1. **Creativity from a novice:** A new engineer can have many interesting new inputs in product design. The lack of experience with the particular product being designed or even with a product development process in general helps with new ideas. A novice has a fresh perspective on the product, as he or she is not mentally conditioned to think about how a product should look and work. While a lot of new ideas proposed by a novice will be shot down, eventually some will survive. New engineers should embrace this learning experience. One caveat is that engineers should be careful implementing new ideas in a design before consulting experienced colleagues. They may not have a complete grasp of the working of the product and may end up modifying it such that other functionalities are negatively affected.
2. **Creativity from an expert:** An expert designer looks at a product in a different way. His or her approach to creativity is a result of years of experience in designing products. Since expert designers better understand the product in question, they are better at suggesting incremental improvements that usually do not affect the functionality of the product negatively in an unrelated attribute. They are also able to access the vast repository of information that they have

accumulated over the years. The author recalls an experience from working in a manufacturing company where an aluminum port had to be removed from a vacuum pump. The port was very critical to the working of the pump, and the method used to remove it had to be such that it did not damage it. To make matters worse, the port was shrunk-fit, so a substantial amount of force was needed to take it out. The fact that it was made of soft aluminum did not help the cause either. One of the more experienced coworkers found a hinged collar with threaded holes on its periphery. The collar was wrapped around the port, and after screws were tightened through the threaded holes, the collar pushed against the port and slowly forced it out. Later it was revealed that the collar had been used previously in somewhat related applications, which gave him the idea. Of course, the knowledge that something like that existed and could be found in a nearby tool bin also helped!

7.3.2.1 **Learning Creativity**

Learned creativity is creativity acquired as one gains experience by watching others. Learned creativity applies to a particular situation or design problem more readily than a generic idea of creativity. An automobile power train designer, for example, can come up with many solutions in the automotive industry, but not so while performing reliability analysis of electronics.

Lack of creativity stems from artificial constraints that the human mind imposes on a solution when faced with a problem. Examples corroborating this can be found while performing mathematical optimization; the algorithms used, by definition, lack these mental blocks. We consider a thought experiment. Consider a computer repository of all available cars and an optimizing algorithm that helps your friend Sheila choose a car most consistent with her preferences. Let us say that Sheila starts by specifying that she prefers a small car. The optimizer outputs a 2-inch toy car. Realizing that she did not specify a minimum bound on the size, she specifies a small car that is big enough for her to fit into, but one that also provides high fuel efficiency. The optimizer outputs a solar car that costs $500,000 and does not work at night! Of course, as she puts more constraints and answers more computer-generated questions, the optimizer better understands her preferences and her trade-off behavior. As a result, an optimal or close to optimal car will eventually be found. One does not expect such erratic solutions when two humans are interacting. This is because we start with many implicit ideas of what a small car is, which constrains our search space. The flipside of starting out with implicit constraints is that new solutions that address the main objectives and not the means objectives become harder to find. For example, let us say that Shiela's main concern was that she was spending too much on gas in her current car. A different, arguably creative solution could be to move closer to work.

A lot of times when optimization problems are set up, key constraints are missing, leading to erroneous results that may not be acceptable at the moment, but reveal that creativity is an artifact of an open-minded search for good solutions. Another important observation to be made here is that creativity is not inherent to humans or even living beings. Many engineering problems have been solved on a computer where the final output surprises designers enough to say, "We didn't know such a solution existed" or "Now we know where to look." Even when creative solutions do not make sense initially, they open doors for new ideas. The fact that the optimizer showed Sheila a solar car might prompt a manufacturer to consider fuel-efficient options in the automobiles they manufacture; something as simple as a rooftop solar panel to run electronics in the car could be explored.

By definition, creativity cannot be achieved solely by following hard rules. To be creative, an engineer must care about the product he or she is designing. He or she must think about its working within the design and manufacturing environment as well as out of it. There are conceptual roadblocks to being creative, as well as theoretical ones. In Chapter 2 we discussed how creating abstractions for the problem at hand could help stimulate creativity. Here we provide some guidelines to help create abstractions/stimulate creativity when we are stuck with a generic solution.

1. Are there fundamental issues that need to be addressed before an improvement can be made to a product? Consider the example of a hard drive in a computer. A hard drive's capacity cannot be increased just by tinkering with the electronics or the mechanical control aspects of it. In fact, discovery of the giant magnetoresistive effect was instrumental in the exponential increase in hard drive capacities. An alternate creative solution could be installing multiple hard drives in the computer, if access speed is not an issue.
2. Am I trying to make an incremental change? Sometimes an incremental change can be accomplished with a minor improvement in the least expected area of a product. For example, if an automobile just fails to meet the legislative fuel economy requirement, the easiest remedy is reduction of weight in nonstructural elements versus making the engine more efficient.
3. Do I understand the working of the product well enough? Many seemingly creative solutions fail when an engineer does not understand the working of the product enough. For example, an engineer may install a smaller, less powerful water pump in a venturi chemical scrubber to meet space constraints and not realize that the pump must provide a minimum amount of head (pressure difference) to create suction in the venturi. Examples exist even outside of engineering design; many times manufacturers dissuade customers

from trying to repair products themselves because a seemingly innocuous and obviously creative solution can cause serious failure or bodily harm. One example could be a customer's replacing the transmission fluid in a car with cheaper motor oil.

4. What is the customer essentially looking for in a product? Many creative solutions are a result of a designer backtracking to the essential function of a product.

7.3.3 Functional Diagrams

A functional diagram is a block diagram representation of a product describing how the different elements of a product (roughly, components) are functionally connected. Functional diagrams do not have specific design implementations for each component. In fact, they help understand the workings of a product so that brainstorming or other concept generation ideas can be used to propose implementations of each component. They are also not unique in that a single product can have many different functional diagrams, but the fundamental similarity between them is that they must address all the customer needs. It is sometimes a good idea to generate different functional diagrams to see if any of them help in generating new implementation ideas.

Functional diagrams decompose the product into its constituent components. The decomposition is functional, and hence the name functional diagram. Each component or set of components in a product that accomplishes a specific task or subtask is a separate entity in a functional diagram. These components interact with each other through "flows." All products accept, generate, or propagate energy. It forms one of the flows within the diagram. Another flow is usually material, for example, staples in a stapler. Components transform or modify energy or material such that they can be used by the components downstream or be outputted. Figure 7.2 shows the functional diagram of a stapler (reproduced from Chapter 2).

7.3.4 Brainstorming for Concept Generation

Brainstorming is a way to come up with creative ideas for a new product design or an existing product improvement using inputs from multiple individuals. Brainstorming involves a setting where a design problem is presented to a group of individuals entrusted with the design of the product, who propose, discuss, and evolve new ideas through constructive interaction. In product design, brainstorming can follow once the customer needs have been identified and a functional diagram is available (though not absolutely necessary). Consider the functional diagram of a stapler in Figure 7.2. For each of the elements of the functional diagram, group members will

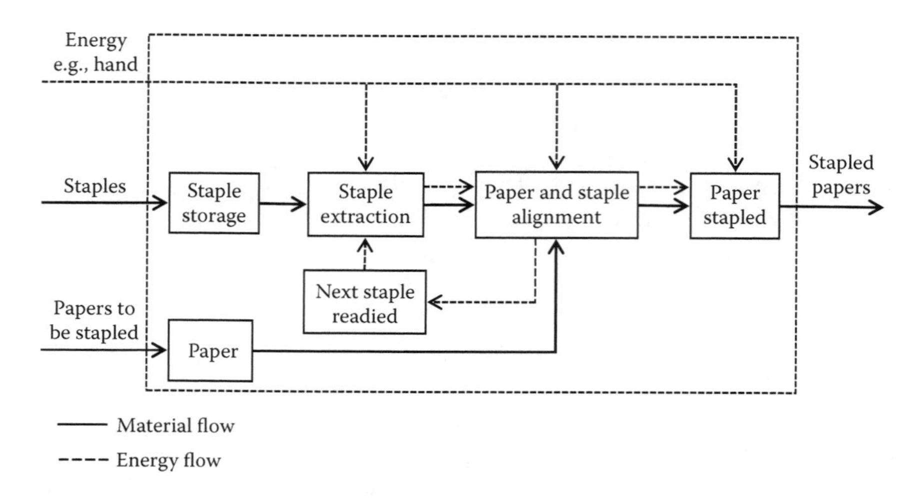

FIGURE 7.2
An example functional diagram for a simple stapler.

propose many implementation solutions. Examples of some ideas for imparting force to a staple are given below:

1. Hand
2. Electricity (mains) and an electric motor
3. Battery and electric motor
4. Chemical charge

Concepts for different components are combined to see if they result in a feasible product. For example, an electrically powered stapler will need provisions for wiring and a way to activate/deactivate the motor, and in addition, it can add new geometric constraints. A hand-powered stapler will need an ergonomic shape for easy application of force. If these requirements conflict with concepts for other components, those combinations should be discarded or at least modified. From the set of feasible product ideas (combinations of component concepts), the best ones can be chosen for further analysis. It is a good idea to generate as many different concepts for each component as possible without regard to how outlandish they are. This is because an idea that seems infeasible at first may become good when improved upon by other group members, particularly in combination with novel implementation in other components. We will see an example of this in the next section, where we cover a chemical scrubber design example. Similarly, a seemingly good idea may be found to be infeasible when more analysis is done. Novel products are usually a result of a fundamental change in how a product is perceived to perform its intended function and how engineering problems are tackled. Creative new solutions must therefore be sought and encouraged during brainstorming.

Brainstorming is now a universally used method in industry to come up with new, ingenious design solutions. Brainstorming, however, has disadvantages of which one should be aware. Group dynamics play an important part in brainstorming. A vocal, confident group member who presents more ideas usually ends up subduing less vocal group members. Fixation on an already proposed idea, called anchoring, is a common occurrence in brainstorming sessions. An idea is proposed that takes over from other competing ideas, and group members simply try to extrapolate from it. Anchoring negatively affects the generation of new ideas, despite all efforts. Therefore, brainstorming should be done in multiple sessions interspersed with times when group members think on their own. If an idea is shot down during a session, the member proposing it may not feel confident enough to propose another. A senior member running the session therefore should encourage input from all the members. Sometimes ideas work in synergy, and therefore an outlandish idea about one aspect of a product may not be as outlandish when something else gets resolved. It is therefore a good idea to keep all proposed ideas under consideration until everyone agrees on a few implementations.

Alternatives/complementary approaches to brainstorming include

1. Independent investigation and thinking on the part of group members
2. Researching implementations of similar products
3. Using stimulus related to the product
4. Using stimulus unrelated to the product
5. Redrawing the functional diagram of the product to see if this gives new ideas
6. Sleeping on it

7.3.5 Chemical Scrubber Design Example

In this section we will implement the steps we learned in the previous sections on a product design example. Specifically, we will go over customer needs encoding and concept generation using a chemical scrubber example. A chemical scrubber is equipment used to render chemical vapors, spent gases, and particulates in chemical processing labs harmless so that they can be released to the environment. We choose this problem because chemical scrubbers are very specific to applications, and almost every order needs to be designed basically from scratch. Consider a company that has just received an order from a customer to manufacture a chemical scrubber. The customer is located in India and requires the scrubber to treat only relatively harmless gases that need to be only "washed" with water. The customer cannot guarantee that there will be continuous water supply for the scrubber, but does guarantee uninterrupted power supply. The amount of gases to be scrubbed is significant, and the experiments do not impart any pressure to the gases, so the scrubber should be able to create suction. The customer also wants the scrubber to provide a visual or audible signal in case of malfunction. In addition, the customer provides a layout of the lab the scrubber will be used in, along with the gases to be scrubbed.

The requirements from the customer need to be translated into clear needs statements so that designs can be proposed. Since the location is in India, the scrubber's electrical system should be able to handle a 220 V power supply at 50 Hz. The amount of gases to be scrubbed implies two things:

1. The scrubber needs to be able to generate suction. The amount of suction or negative pressure needed is the function of plant layout (which determines the distance from sources) and amount of gases to be scrubbed.
2. The scrubber should be able to treat the amount of gases released. This determines the size of the scrubber, amount of water circulating, as well as surface area of contact between gases and water.

Scrubber design depends a lot on chemical calculations; therefore, the amount and quantity of all the gases need to be known. Also, the layout of the laboratory in which the scrubber needs to be installed determines the negative pressure that needs to be generated. We omit the calculations here and assume that 5 inches (water column) of suction pressure will be enough (which includes a factor of safety) to draw the gases to the scrubber. Assume also that it is determined that a packed tower is needed to scrub the amounts and types of gases encountered. A packed tower is essentially a vertical pipe filled with polyurethane or plastic balls that increase the surface area over which the flue gases and water have to interact. Using this information, the following needs statements are generated, which make the customer requirements more concrete.

1. The scrubber needs to generate suction pressure of 5 inches of water for incoming gases.
2. The scrubber needs to have a packed tower for scrubbing.
3. The scrubber cannot rely on continuous water supply.
4. The scrubber should have warning alarms in case of malfunction.
5. The scrubber should accept a 220 V power supply at 50 Hz.

Notice the way the sentences are framed. They focus only on the requirements of the scrubber, are mostly independent of each other. and do not provide so much information that they imply a particular design. Customer needs statements should be shown to customers before concept generation begins to ensure that nothing of importance is missed.

The next step of concept generation is the creation of a rough description of the new product, which helps in coming up with some candidate designs. A lot of information collection and soul searching happens at this step. The input to the concept generation phase is the description of what is required of the product (e.g., the needs statements above), and the output is a set of candidate designs (not just one) that can potentially fulfill these requirements. The candidate designs are then refined/evaluated and the best ones selected. Notice that brainstorming, which we discussed earlier, forms only a part of the concept generation process.

As we have mentioned earlier, a decomposition-based approach to concept generation is advocated in product design. This is done primarily because most engineering problems are too complex to solve in a single step. The design problem is divided up into multiple subproblems, and the subproblems in turn are tackled separately. Ulrich and Eppinger (1999) present a five-step approach to concept generation. In this section we will present a similar approach for the scrubber design example. The design group that has been entrusted with the project meets to discuss the deliverables. It is common for different group members to initially have different ideas (or no idea) as to what needs to be done. The first meeting, usually termed the kickoff meeting, accomplishes exactly that. It brings all the group members onto the same page in terms of problem definition. There is also a division of responsibilities among different group members. Brainstorming for different design concepts can start in the kickoff meeting, but generally no design decisions are made.

Research reveals that a chemical scrubber needs to be designed for the particular gases it is supposed to treat. Scrubbers that work with silane, for example, may not be very good for working with ammonia, and vice versa. Moreover, certain chemicals are too corrosive for simple scrubbing and may need oxidizing and even a catalyst. A quick call to the customer will reveal what gases will be present. Recall that in this example the gases are relatively inert and can be scrubbed easily just with circulating water. It is important here to start creating a functional diagram. Figure 7.3 shows one way the functional diagram for the chemical scrubber could be drawn. Recall also that while making functional diagrams, it is essential not to assume any particular implementation of a subproblem. For example, saying "connect ½-inch water inlet line" will preclude situations where continuous water supply is not available and the lab employees use buckets to fill the scrubber reservoir.

The next step in concept generation is to collect knowledge regarding what concepts exist or have already been implemented. Sometimes information can be found in-house from previous projects, but generally for new products a lot of external research is required. This might include reverse engineering already existing designs, asking industry experts, or performing a literature survey. Rarely does a firm encounter an engineering challenge that has never been tackled. Problem decomposition helps tremendously in a literature survey, because it is easier to find solutions to specific problems such as how to generate a negative pressure of 5 inches of water, as opposed to the general problem of how to design an optimal scrubber with a given set of specifications. A literature survey should not be restricted to only easily accessible online journal articles. Product manuals, textbooks, and occasionally patent filings also prove to be very helpful.

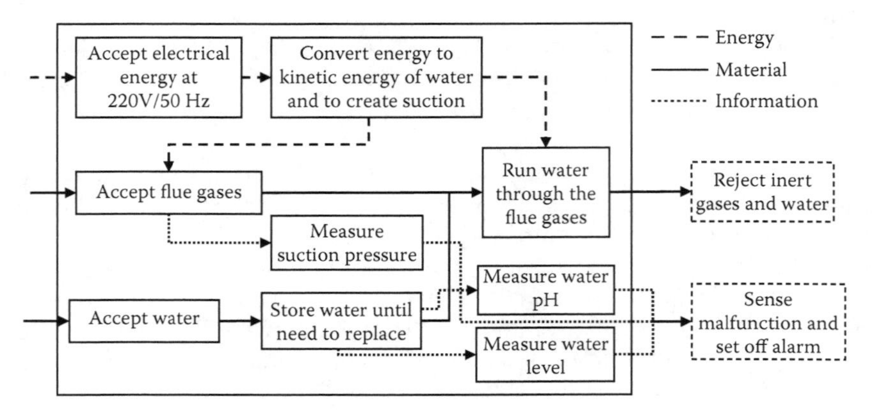

FIGURE 7.3
Functional diagram of the chemical scrubber.

We now come to the fun part of actually proposing concrete solutions to the subproblems, usually done in a brainstorming session. Accepting a 220 V/50 Hz power input is straightforward; care should still be taken that all the components within the scrubber's electrical systems are compatible with it. Since the scrubber is going to be used in a different country (India), one must look at local electrical standards. Different countries have different electrical safety requirements, and they must be adhered to. Converting energy has two aspects, each with multiple solutions. Notice that at the concept generation stage we do not limit the different proposed solutions, regardless of how outlandish they may be.

1. Convert electrical energy to kinetic energy of water:
   a. Use a water pump.
      i. Use a centrifugal pump.
      ii. Use a vane pump.
      iii. Use peristaltic pump.
   b. Evaporate water by heating and let steam flow through the flue gases.
   c. Do not use electric power to move water. Store water above the scrubber and let gravity flow the water through the flue gases (continuous refilling of the reservoir will be required).

2. Convert electrical energy to create suction for flue gases. This can be done in many ways as well:
   a. Assume flue gases will flow into the scrubber anyway; that is, the customer's equipment provides the positive pressure.
   b. Use an air pump.

c. Use a fan.
d. Use a venturi.
e. Cool the inside tubes enough so that a negative pressure is generated.

Similarly, running water through the flue gases can be accomplished in many ways. One can spray water through nozzles in the pipe carrying the gases. Recall that the customer essentially asked for packed tower, which is a vertical pipe filled with polyurethane or plastic fillings that increase the surface area over which the flue gases and water have to interact. Of course, the size and type of packing need to be determined, but such issues are generally not considered in the concept generation phase. Water storage can be in the form of a tank installed within the scrubber. A water tank is a good idea even when there is a continuous water supply because it can allow for maintenance of water lines; that is, the scrubber can be run from the water in the tank while maintenance is taking place.

Now since a packed tower is required, there needs to be a way to continuously pump water within the scrubber since there is no continuously running water supply at the installation location. This can be accomplished using an electrically powered off-the-shelf water pump. The pump needs to be sized (horsepower) based on this requirement and needs to be able to run at 220 V/50 Hz. If we look at another customer need of creating suction for incoming flue gases, we see that instead of using an air pump, a venturi achieves the same purpose. Also, since the scrubber already has a water pump, the venturi can be powered by it. This way three customer needs (creating suction, running water through packed tower, and not relying on continuous water supply) are satisfied. Moreover, the venturi provides other advantages, for example, scrubbing action resulting in reduced packed tower size, as well as the ability to handle particulates that most air pumps cannot.

Adding venturi, however, increases the required power rating of the water pump within the scrubber because it now needs to create enough suction to pull in the flue gases. Figure 7.4 shows the schematic of a standard venturi used in chemical scrubbers. The venturi is usually made of PVC because of its relative inertness to chemicals. A pipe-carrying high-pressure water runs through the top cap of the venturi. This water is then sprayed into the throat (decreased cross-sectional area) of the venturi using a nozzle. The nozzle is designed to impart turbulence to the sprayed water. Sometimes this is accomplished using a metal insert in the nozzle to increase turbulence. The water spray creates enough negative pressure to pull flue gases from the left, higher-diameter port. Venturi suction is very sensitive to the height of the nozzle within the throat, head generated by the pump as well as the velocity/turbulence of the spray. A pressure gauge on the inlet is used to try different water pressure levels, the height of the nozzle as well as the metal insert in

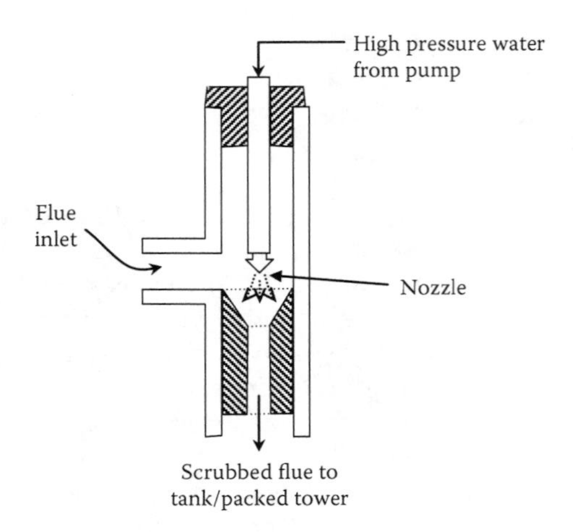

FIGURE 7.4
Schematic of a venturi used in a chemical scrubber.

the nozzle. Once a reliable and stable suction pressure of 5 inches is established, the configuration is noted (including the size of the pump that will generate the required pull). A company that anticipates making many scrubbers in the future would characterize the performance of the venturi over a wide range of design parameters. Figure 7.5 shows the schematic of the entire chemical scrubber.

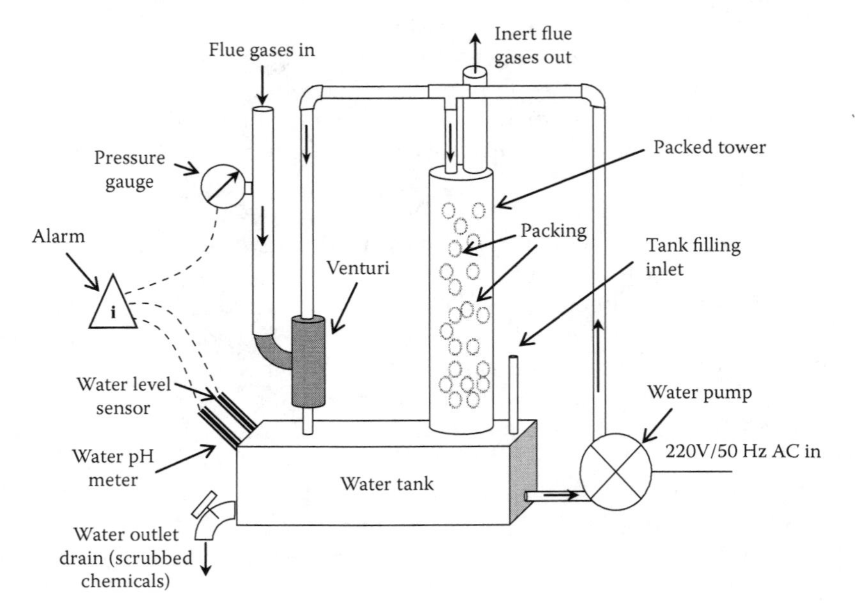

FIGURE 7.5
Schematic of the chemical scrubber.

7.3.6 More on Concept Evaluation

Product performance for each concept proposed is not always known with certainty. While experience and common sense can help eliminate most inferior concepts, one cannot be sure that even the ones that do not seem to have any obvious shortcomings will not fail when a prototype is built. Consequently, it becomes hard to correctly evaluate concepts. These are the reasons why in Chapter 2 we showed the meta-decision one makes in the very early stages of concept selection to select the designs that promise high expected utility from profits. As concepts get more and more concretized, there is more information to be used. In our scrubber example we saw how different combinations of solutions to the design subproblems can be analyzed such that a good combination (nearly optimal) solution is reached. For a relatively simple product such as the chemical scrubber we just discussed, this is pretty straightforward. For complex products, though, we need to discuss some methods that are commonly used.

One of the easiest ways to evaluate and select the best concept from a set of candidate designs is to ask the customer. For a clear order to build a product, the manufacturer shows the potential designs, including their pros and cons, to the customer, who can select one or two, which can be further refined. In cases where there is no clear order and the product is being made for the general public at large, lead users can still be asked to evaluate and provide inputs on different candidate designs. Software companies take this route regularly by first releasing beta versions of their products and collecting end user feedback. Many times, because of various reasons, including intellectual property protection, end user inputs cannot be acquired. In such cases, a company needs an internal methodology to evaluate different concepts.

Most of the time, while internally evaluating concepts, experience and intuition play an important part. Designers analyze the different designs, think about the working of the product based on the implementation of each design, and select the best one based on judgment and experience. A formal approach, however, is still recommended here. Ideally, one would find the expected utility from each design based on the distributions of revenue and cost (and hence profit). If the manufacturer does not feel comfortable assigning a probability distribution to the profit from each concept, the following approach can be taken. For each of the following items, a score between 1 and 10 can be assigned to each candidate design.

1. Assess the attributes of the product for each implementation, for example, power requirements, cubic feet of gases scrubbed every hour. Determine how well each meets the requirements.
2. Determine the cost of manufacture and expected profit. Take into account possible warranty issues.

3. Think about safety issues. Are certain designs safer than others?
4. Think about reliability issues. Is a particular design more susceptible to failure than others?
5. Think about manufacturing. Which design will be easier to manufacture?
6. Parts for which design will be easier to procure?
7. Which design is going to be more appealing based on form?
8. Any other product-specific metric.

For each of the candidate designs one can then find a final score by finding the weighted sum of scores for each item on the list. Of course, some items in the list are more important than others (e.g., ability to meet attribute requirements is more important than the product's appeal based on form). The relative importance is reflected in the weight associated with an item, which can be arrived at based on discussion among the design team members. Thinking in a systematic way like this ensures that all the important design issues are considered and a numeric metric is used that takes away biases. Notice that while the simplistic approach is not entirely normative, it does approximate a value function. Furthermore, thinking along the lines of the eight points given above can actually help assess revenue and cost distributions.

7.4 Some Concepts in Product Development

We devote this section to some key concepts in product development. While they are not presented in any particular order, familiarizing oneself with these ideas will go a long way toward understanding the breadth of the field.

7.4.1 Axiomatic Design

Each product and each design is different. As a result, it is hard to come up with a predefined set of rules that apply to all products and all possible designs. In the context of ground rules that must be followed, we consider two axioms that apply to engineering design as proposed by Suh (2001):

1. Maintain the independence of functional requirements.
2. Minimize the information content of the design.

The above two axioms hold in that order. Properly interpreted, they can be stated, "Generate designs that maintain independence of functional requirements, and of all the designs that do that, select the one that requires the least number of instructions to produce." We now discuss the key phrases in the axioms in detail.

7.4.1.1 Maintaining Independence of Functional Requirements

Functional requirements address the intended behavior of a product during operation. A product may have many functional requirements, usually depending on product complexity. Consider a standard office stapler, for example. A stapler has to satisfy at least three functional requirements: (1) impart enough force to the staple that it pierces a reasonable thickness of paper, (2) bend the staple once it reaches the other side of the paper for safety and to keep the staple from falling out, and (3) ready the next staple for the next round of stapling. Now consider an electric stapler that imparts the force using an electromagnet instead of a human operator. Clearly, this allows for a properly applied and repeatable amount of force. Furthermore, if large forces are required to staple thick bundles of paper, the strength of the electromagnet can be easily increased. Now consider the scenario in which incorporating the electromagnet affects the ability to ready the next staple. This will require redesign of the whole stapler, leading to increased costs. A stapler that started out with a design that had minimum dependence between the stapling action and readying the next stapler may need minimal modification when the type of force input changes.

Thurston (2001) points out an important misconception in design literature. Independence conditions for functional requirements are often confused with independence conditions noted in a multiattribute utility function assessment (see Chapter 3). The two are fundamentally different. Two attributes (as determined by functional requirements) can be probabilistically correlated or even perfectly dependent on each other but still be preferentially and utility independent. Preferential and utility conditions have to do with independence of attributes as perceived by the decision maker, and as such do not depend on probabilistic dependence between them.

7.4.1.2 Minimizing Information Content

Information content of a design roughly translates into how much information is required to completely manufacture it. A design with higher information content adds manufacturing costs, increases redesign efforts, and leads to material procurement, as having well as having other logistical issues. There are many metrics that can be defined and can act as surrogates for information content of a design:

1. **Number of parts:** Number of different parts in a product, or part count, is a strong indicator of information content of a design. Unless parts share a lot of commonality (e.g., two dozen screws), they add a tremendous amount of variability to the manufacturing process.
2. **Number of assembly steps required:** Number of assembly steps is a measure of time of assembly that directly translates to labor costs. Furthermore, a large number of assembly steps also increases the probability of error and, thus, rework.
3. **Number of different material types:** Different materials in a design means that raw materials from potentially different sources need to be ordered, stored in inventory, and then processed. Different materials also have different properties and cannot be processed using the same methods. Material mix also reduces the ability to recycle certain products.
4. **Number of fasteners required:** Too many fasteners not only increase assembly time, but also reduce the overall integrity of the product. Alternative fastening methods, such as gluing, welding, and snap-on tabs, should be explored where possible.
5. **Average assembly time:** Average assembly time roughly takes into account the above metrics and can be a measure of information content.
6. **Size of the CAD file:** A lot of design engineers interact by sending a CAD file back and forth while collaboratively working on a design. It is considered the blueprint of the design. The size of the CAD file is sometimes used as a surrogate for how complex a design is.
7. **Number of machining operations:** Different operations require different machines, setup, orientation, and specially trained workers.
8. **Number of different dimensions used:** If a product uses many different dimensions for similar-looking parts, it is going to be difficult to manufacture.

It has been seen in practice that information content of a design is a very good measure of the cost of the final product. This is especially true for labor-intensive products. Simple designs also tend to be flexible; they are easy to evolve as technology improves and new functionalities are added. Each of the surrogates above individually does not tell the whole story, but generally taken together, one arrives at the broad picture of the information content of a design.

7.4.1.3 Conflict between Axioms

It is clear that axioms potentially conflict. Consider the front fascia of an automobile that has integrated fog and turn lamps. Making the fascia in one injection molded piece reduces the number of parts and can even reduce assembly time. Furthermore, since fog and turn lamps are in a single assembly, their manufacture, assembly, and wiring are also simpler (minimum information). However, provisions will need to be made to allow for bulb replacement without the removal of the fascia. Also, a minor accident that damages the fender will affect the lamp assembly as well (lack of independence). Therefore, independence of functional requirements can and does interfere with the minimum information content axiom. The key is to understand that independence of functional requirements generally takes precedence over the minimum information axiom.

7.4.2 Role of Metrics

Metrics are mathematical constructs (usually a number) that can tersely convey a piece of information. This information in general will be harder to convey succinctly otherwise. For example, while driving, your speed over the prescribed speed limit is used as a metric for your likelihood of causing accidents. You may be able to prove, based on the car you drive or your driving history, that you are no more likely to cause an accident when you speed, but you will still get a speeding ticket. This is because metrics standardize different situations. They also help an external agency evaluate a situation. Metrics are used routinely in engineering design and product development in particular. Metrics help standardize an amorphous process such as product development by giving it structure.

We have been (and will be) discussing metrics throughout this chapter. Part counts, number of assembly steps, etc., are all metrics to determine the information content of a design. Metrics such as percent job completion and money committed are used to judge the progress of a product development process. Reliability is used as a surrogate for quality and so on. It should be noted that while metrics are very useful, they should be used carefully with their shortcomings fully known and understood. For example, part count can be used as a surrogate for the cost of a design. However, it assumes that there is a deterministic relationship between the two. It is possible that the cheapest and most reliable design has a large number of different parts.

7.4.3 Product Architecture

Product architecture is the physical arrangement of components within a product. Two determinants of a product's architecture are design requirements and softer engineering (e.g., aesthetics) requirements. Design requirements (including assembly and compatibility) take precedence over aesthetics. Even after a product design is finalized and the components have been selected, they still need to be physically arranged within the product. In this sense, product architecture can be classified into three types: bus, slot, and sectional architectures (Ulrich and Eppinger, 1999). A bus architecture incorporates different components of a product in a more or less parallel fashion. Parallelism here should not be confused with redundancy because the components usually perform different functions. The components are added to a central component that supports them and coordinates their functioning. The system can generally be updated, modified, or repaired easily because there is limited interference among the components. The best example of a product with a parallel architecture is a personal computer. In a personal computer, different components, such as video card, processor, and memory, attach to the motherboard that powers them and coordinates their performance, even providing a place for physical installation. Slot architecture (the most common type) works when components cannot be interchanged. Not only are the interfaces with the rest of the product components different, but they also perform entirely different functions. A much less commonly encountered architecture is sectional architecture, where interfaces match (as in, for example, floor tiles), but the connection is progressive in that there is no central component.

7.4.4 Softer Engineering Requirements

Softer engineering requirements get their name because while they do not directly affect the ability of the product to perform its function, they do enhance its appeal to the customers. As a result, the manufacturer gains by getting a better price margin, better reviews, and even reduced warranty claims. We consider two commonly considered soft requirements, aesthetics and ergonomics.

For most technological products, function drives form. At the same time, once technology matures, form acts more and more as a differentiating factor. This can be verified from the evolution of many products. Automobiles used to be merely a collection of parts that worked together to move the vehicle. Today automobiles are supposed to look sleek and offer many different colors, a grille, fascia, and comfort features. Automotive companies operate under many different names to offer slight variety to customers. Similarly, some electronics companies thrive on appearance and aesthetics. Making a minor investment in making a product distinct in appearance helps customers better relate to a product. Certain finishes, for example, the matte aluminum look in electronics, have been found to have better aesthetic appeal.

Product ergonomics determine a product's ease of use for the end user. A product that requires too much physical strength or dexterity is not a well-designed product. The amount of effort that goes into designing an ergonomic product is a function of direct human interaction with it. A car seat needs a more ergonomic design than a wall clock. Products where ergonomics play an important role must be evaluated, if possible, by the end users. Every step taken in operating the product should be carefully monitored and avenues for improvement identified. Safety is a closely related concept to ergonomics. Sharp corners, unnecessary vibrations, risk of electrocution, etc., should be minimized in ergonomic product design.

7.4.5 Product Pricing: Cost-Plus and Target Costing

A product needs to be priced in such a way that it maximizes profit margins and at the same time does not drive away the customer. Product pricing is the process of determining what the final price to the customer is going to be, given a profit margin *or* how much money should be expended in making the product given the profit margin. This is easier said than done. For most products initial development cost is large. As a result, it is hard to determine the unit price of the product to the customer because not only is the demand uncertain, but one does not know how long the product can be successfully marketed. On the other hand, if we wait too long to resolve these uncertainties, we might have spent so much money that the product cannot be profitably marketed.

There are two techniques that are commonly used:

1. **Cost-plus method:** The cost-plus method dictates that the price of a product should be the cost of manufacture of the product plus a profit margin. The approach is simple and intuitive, which is why it is so prevalent. However, this approach is by definition oblivious to what the customer expects to pay for the product. The cost-plus method has been the traditional method of choice for most manufacturers. If profit margins are running thin, to reduce cost, one mainly focuses on waste reduction and efficient operations. Once issues arise, it is very hard to pinpoint the actual cause of reduced profit margins. A manufacturer using the cost-plus method is ill-placed to counter competitors.
2. **Target costing method:** In target costing, the price of the product is determined in the concept selection stage itself. This pricing approach requires a good understanding of what the customer is willing to pay and whether such a product can be profitably manufactured. The price, minus the profit margin, is cascaded down to the individual operations. Target costing allows for a tight control over where all the expenses occur. Furthermore, understanding what the product should cost, a manufacturer also is well placed to ward off competition. There is a sharp geographical divide when it comes to how products are priced in the United States and in Japan. The majority of the companies in the United States still use the cost-plus method, whereas almost every manufacturer in Japan uses the target costing method.

Needless to say, target costing should be the method of choice. While cascading a product's final price to individual operations, one cannot, however, assign costs to every single operation. Eventually one comes down to a minute operation where assigning cost will be too time-consuming or even detrimental. In such cases a cost-plus approach can be used. For example, consider a machine that requires 30 different welding operations. It makes sense to assign a total time to the welding operations (and hence cost) versus allocating times to each welding operation.

7.4.6 Logistics

Logistics influence all engineering decisions. Logistics are all the extraneous factors (mostly outside of technical) that affect the fulfillment of a service or a product. The job of a company is far from complete with the design and manufacture of a product. The company needs to procure all the licenses, line up suppliers, figure out transportation, manage inventory, decide placement within stores, manage incentives to retailers, and so forth. Also, if and when we start considering recovery of products and materials after use, we have to consider reverse supply chains as well. There is legislation all around the world, especially for electronics products, that the manufacturer is responsible for the ultimate fate of the product. For example, the waste electrical and electronic equipment (WEEE) directives in Europe have mandated a recovery target by weight for electronic products. Clearly logistics issues should be considered with regards to the following:

1. Time and money requirements in material procurement
2. Lead time in various manufacturing operations, including procurement delay
3. Inventory management
4. External factors impacting product cost
5. Transportation costs and delays
6. Legal and intellectual property issues
7. Negotiations with material suppliers as well as product retailers
8. Legislative mandates affecting design, manufacturing, and sale
9. Effect of uncertainty associated with each of the above issues

7.4.7 Market Size Assessment

Market size assessment helps predict expected revenue from marketing a product. Expected market size therefore determines feasibility of offering a new product. A simplistic bottom-up approach to market size approximation involves knowing the total population size, P, and then simply taking into account the fraction of customers who would be interested in the specific product, F. In a limited geographic area, the expected market size, M, is given by

$$M = P \times F \tag{7.1}$$

The fraction F can be broken down into various factors. Indeed, there are different ways to partition F, but the general approach involves the following: How many individuals use the product or a similar product? How many are aware that the new product is being offered? How many will be interested in switching to the new product? What is the effect of advertisement? What is the effect of word of mouth? Even using this simplistic approach, many strategies can be created to increase the market size. For example, advertisements can help increase awareness about the product, and knowing what attributes are valued by customers and making required modifications can increase the switchover rate. Market size forecasting is a complex science, and substantial effort is required to make the predictions accurately, particularly when multiple products are offered and in the presence of competitors. In Chapter 8 we will cover some commonly used methods to forecast demand of a product or a set of products.

7.4.8 Mass Customization

Most concepts we learned in this chapter can benefit from or enable mass production and realizing economies of scale. When a company streamlines a product development and manufacturing process, its per product expenditure goes down since the relative influence of fixed capital cost decreases. Cycle times become predictable, and so do rework and maintenance. However, a customer trend toward mass *customization*, which allows each customer to specify a product configuration, works antithetically to mass *production*. For certain products, such as personal computers, customers require and even demand the ability to customize the products tailored to their needs. While initially this sounds as if it can decrease profitability, mass customization has been shown to lead to competitive advantage in many situations. Mass customization requires strategizing on the company's part. It must be cognizant of the often cited 80-20 rule, which says that almost 80% of the profits are realized by 20% of the product offerings. This means that a company should still focus its energy and efforts on mass-produced products. The customizable products should bring in a substantially higher price premium. Also, the customization should be in superficial parts that make different products "appear" different without requiring fundamental changes in design or manufacture. In Chapter 2 we covered component and process commonality within a product family. Mass customization should always include commonality as a consideration.

7.5 Design for X

Decision based design can be very helpful when there is a specific focus added to the product design process. For example, imagine an automobile manufacturer that is considering designing a driver-side door. Aside from the general requirements of opening and closing mechanisms, strength, size, and shape, it can consider factors such as:

1. How easy is it going to be to manufacture?
2. What are the assembly and disassembly steps?
3. How much environmental impact is generated?
4. Can the parts be standardized across different cars?
5. Can the door be serviced easily?
6. Can the door be replaced easily?

As is evident, different designs can result when the focus is shifted. A door designed for easy assembly may be hard to manufacture because of additional constraints imposed on the designer. Considering some general guidelines is therefore important so that a design does well on all of the above parameters.

7.5.1 Design for Manufacture

As the name suggests, design for manufacture, sometimes abbreviated DFM, involves actively considering manufacturability of a product in the planning and design phases. While it is well known that manufacturing is generally the longest and most labor-intensive activity in product development, most of the manufacturing time and cost is determined in the design phase itself. As a result, improvement in manufacturing methods is going to have less of an effect on the whole cycle time. Instead, thinking actively about manufacture during design and being able to foresee issues help create a design that is substantially easy to manufacture and has other significant benefits, such as reduced cost and reduced rework, resulting in increased profit margins and predictable cycle times.

Some examples of design for manufacture actively considered in design are as follows:

1. Specifying the right materials in terms of both cost and workability
2. Specifying exact operations to be performed for manufacture, thereby being able to predict manufacturing cost
3. Providing exact specifications and tolerances
4. Using similar processes for similar parts
5. Minimizing part count through parts integration, if possible
6. Minimizing the number of different operations
7. Standardizing components to be used

Design for manufacturing can also be broken down into various subdisciplines of design for assembly, design for disassembly, design for modularity, and so on. In the following we discuss the subdisciplines separately.

7.5.2 Design for Assembly and Disassembly

One of the ways to reduce the overall production cost is to reduce the assembly time and effort. This translates into also looking at designing for disassembly, which becomes important in the context of repair or remanufacturing of a product. Boothroyd and Dewhurst's (1990) method has been shown to substantially reduce assembly–disassembly cost of products. The method looks at every detail of the operations and presents guidelines, some of which are

1. The components should be insertable from top. Since gravity acts as a stabilizing factor, assembly is more likely to be finished sooner. Moreover, a top-down assembly method gives the operator a clear view of the assembly so far, which can be incrementally assembled if all the parts can be inserted vertically.
2. The components should be such that there is a natural way for them to be assembled. Rails and matching features in the component to be inserted can simplify assembly. Similarly, while a square threaded nut is hard to assemble, acme threads are easy to assemble and accomplish almost the same power requirements.
3. The components should require a minimal number of tools and should be hand assemblable, if possible. For example, snap-on tabs are very useful in plastic components since they do not require any tightening.
4. The components should not require complicated motion to assemble, and once assembled, they should be secure.

Design methods aimed at assembly also help with disassembly. Disassembly may be required for repair or for reusing some components in next-generation products. A well-documented assembly process can allow for the steps that need to be taken in reverse order for disassembly. Products that are candidates for reuse must consider disassembly steps carefully, and to this effect, many disassembly planning methods, such as the reverse-fishbone diagram and incidence matrix methods, have been developed.

7.5.3 Design for Modularity

Design for modularity can be a subset of design for manufacture, but in distributed design it takes a totally different flavor. A modular design is one where different subsystems within the product not only are independent in terms of design and function, but also can be independently installed and replaced. A key driver of modularity is the design of interfaces. A personal computer is an excellent example of a modular product since different components can be installed and upgraded almost independently. This is because components mainly share information and not actual physical quantities such as material or force. Modularity can still be achieved in other products, for example, in piping systems, by using standardized interfaces. Most mature products have evolved to be very modular, for example, automobiles, airplanes, staplers, and military weapon systems.

Modularity is closely related to the independence of functional requirements axiom. In distributed design or design that takes place with many designers separated by vast distances, modularity becomes even more critical. Interfaces between different components are predecided and form hard constraints for each designer. Consequently, when a manufacturer decides to outsource some components, exact specifications must be specified and met in terms of dimensions and interfaces.

7.5.4 Design for Environment

Environmental considerations are becoming more and more important in the context of today's economy. Aside from increased customer awareness, many countries have laws that put stringent limits on the environmental impacts during manufacture as well as during use of products. Sometimes products even need to be designed for reuse so as to keep them from going into the landfill. Manufacturers also use some green technologies for the "halo effect," or creating the feeling among customers that they are environmentally responsible, while most of their products may not be environmentally benign. These put additional constraints on the product design and development process. In this section we will briefly touch upon different strategies that manufacturers consider for reducing environmental impacts of the products they manufacture.

Environmentally benign manufacturing entails reducing the environmental impacts created as the product is being manufactured. Energy-intensive operations such as welding and face milling can be minimized. This might require starting with blanks that are close to the desired dimensions. Some materials are harder to work with than others, and designs should not recommend these unless absolutely necessary. Finally, some processes, particularly in electronics manufacture, release toxic chemicals into the atmosphere and the groundwater. Such processes should be minimized or eliminated, and if this is not possible, there should be measures in place to properly treat the toxic waste. The total environmental impact from a product over its lifetime can also be minimized by making it environmentally benign during use. Automobiles, for example, cause most of their environmental impacts during their use. Therefore, a fuel-efficient car can have less of a negative total impact on the environment even if it causes more environmental impacts during manufacture.

Many times environmental impacts can be mitigated by simply reusing components from old products. This method has the advantage that even products that were not manufactured in an environmentally benign way can be kept from going to the landfill (also avoiding manufacture of their replacements).

This particular implementation of environmentally conscious design is dubbed *design for remanufacture and reuse*. Modular products lend themselves well to component-level reuse, when full reuse is not possible. Mangun and Thurston (2002) even showed with a personal computer example that if one makes component-level *optimal* decisions of whether to reuse, remanufacture, recycle, or discard each component, not only can one get a product that is environmentally benign, but one can also realize significant profits for the manufacturer by recovering value in the products that would otherwise have been lost. Design for remanufacture and reuse is an active area of research.

7.5.5 Design for Reliability and Service

Design for reliability aims to design and produce products that will fulfill their functional requirements for a prespecified amount of time, with a high probability. Customers use products for a variety of different tasks. Failure of products during use can cause anywhere from a minor inconvenience to major injuries or loss of life. Clearly, these occurrences have to be avoided since the resulting loss of reputation or lawsuits can substantially affect a company's future prospects. Even in terms of warranty costs, product reliability or the lack of it has tremendous implications. For example, warranty costs, if failure occurs, are comparable to the profits for many automotive companies. Design for reliability and design for service go hand in hand. A reliable product needs service less frequently. Moreover, a product designed for service will incur lower repair and servicing costs. As expected, modular design with proper attention to disassemblability further drives service costs down.

Design for reliability is a subset of reliability engineering. Very often it is hard to know a priori if a particular design is going to be reliable, particularly considering uncertain operating conditions. Therefore, experience from the previous generation of products is often used to identify the dominant failure modes of a product. Such failure modes are either mitigated or eliminated from the next generation of product. For many products, reliability (or quality) acts as a competitive advantage in the marketplace. Companies that sell reliable products can realize significant profit margins over their competitors, further improving their returns from investing in reliability. Japanese automakers gaining large market share in the United States is a classical example of reliability acting as a competitive advantage.

7.6 Technology Adoption

A critical element of product design is the time content. Proper timing of minor or major updates in a product is critical, as there are two competing forces: (1) customers routinely want products that are sufficiently improved from their predecessors, and at the same time, (2) they do not want the products to be so different that they become hard to use, and the customers need to retrain themselves. Even a company with a successful product enjoying a substantial amount of market share needs to constantly evolve its products to stay ahead of the competition. Marketing literature is replete with case studies where such evolution was done successfully and, in many cases, unsuccessfully. As such, there is no general recipe that is applicable for all products and all markets, but basic rules exist.

New technology generally follows the well-known S-curve. Figure 7.6 shows an example S-curve. Once a new technology is discovered, there is inertia to be overcome. This inertia comes from many challenges, which might include overcoming research challenges, procuring resources such as new materials, creating trained manpower, and creating customer awareness about new products, among others. As a result, initially a slow rate of improvement is seen. As understanding of the technology improves, there is a steep rate of improvement in its application and acceptance. Finally, a time is reached when the technology is fully mature and all the possible improvements have more or less been made. At that time, the curve flattens out and the slope becomes horizontal. Additional improvements in the technology cannot be easily accomplished. Furthermore, as competing technologies appear, time and energy are diverted to these new technologies (the second curve), leaving fewer and fewer resources for the old, already mature technologies. Notice that the newer technology lags behind the established technology for quite some time before research and development makes it better. It is hard at this stage to predict that the new technology will indeed take over. This is one of the reasons managing advanced technology is difficult.

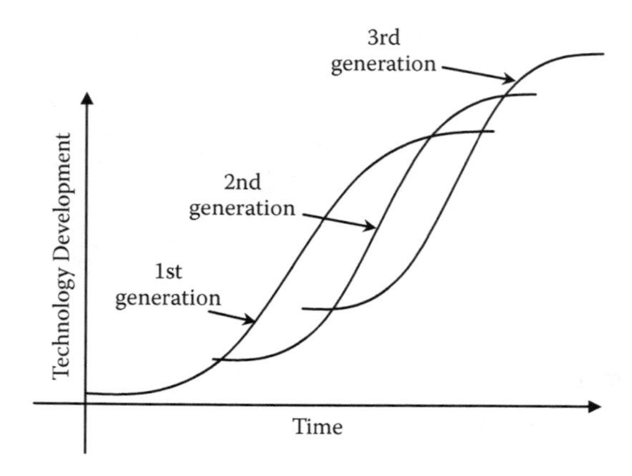

FIGURE 7.6
The technological S-curve. Newer technologies start lower than established ones, but research and development usually allows them to overtake.

Historically, many technologies have been seen to follow the S-curve. Computer hard drive capacities, disc brakes in cars, digital cameras, and internal combustion engines are a few examples. All these technologies saw a period where there were many technical challenges that needed to be overcome; as technologies matured, there was better acceptance and marketplace competition that led to increased research and improvements in the products. The technologies have more or less matured, and the differentiating attributes are few. Understandably, the technological S-curve is closely followed by the market adoption curve (for example, as in the Bass diffusion model, which we will discuss in Chapter 8). This is no coincidence because the market proves to be the driver for technological improvements.

7.6.1 Managing Advanced Technology

Managing advanced technology is always a challenge. A firm has to set aside significant amounts of assets and manpower resources to make the new technology a reality, the monetary gains from which are, at best, uncertain. The first decision a company makes is the make–buy decision. A make–buy decision determines whether a new technology is worth developing in-house or should be outsourced or licensed from another company that has already developed it or can do it more cheaply. There are pros and cons of both. If developing new technologies in the field is the core competence of the company, it should develop the technology in-house. Otherwise, if the company engages only in marketing the technology or if the new technology is just a differentiating add-on to the product, the company is better off outsourcing.

Automotive companies in general are justifiably called assemblers these days. Contrary to the times of Henry Ford, they outsource a substantial fraction of their components. If, for example, a car company does not traditionally develop antilock brake system (ABS) control modules, there is little incentive to develop the component when the need arises. The amount of effort it would take to go out of the core competence of the company is barely justified. Electronics companies are always going to be better equipped to design new and more reliable modules. If there is a new technology to be adopted in the electronics realm, they will also be set up for it. Furthermore, since ABS units form a small component within a car, and are rarely a differentiating feature, there is little risk of losing competitive advantage. Clearly, outsourcing decisions can also have disadvantages. Companies that do outsource risk losing market share because over time they lose the skill to make competent products. They also end up paying huge premiums to the companies or suppliers to whom they outsource. Increased dependence on the suppliers also adds to lead time and uncertainty in component availability and compatibility. Finally, warranty claims are complicated when components are manufactured by many different manufacturers, as they still have to be managed by the assembler.

If it is decided to make the component in-house, the company needs to manage its design, production, integration, and marketing. The strategies differ significantly between large and small companies. In the following we talk about them separately.

7.6.1.1 Managing Technology in a Small Company

A small company faces higher uncertainties. A new technology can be the difference between high unexpected returns and even possible bankruptcy. However, the company has the flexibility to incorporate an additional endeavor and less to lose in terms of reputation. A small company that has flat structure is further suited to develop and manage new technology. There is a better division of labor, and the effects of developing and marketing new technology on other company departments are generally known. As with many other product development concepts, managing technology in a small company can be difficult or easy, depending on the industry. In the medical implant industry, for example, many small companies exist in niche markets with new products launched and patented frequently. They maintain close contact with their customers (hospitals and doctors) to ensure that the market needs are immediately addressed. Conversely, it will be very hard for a small company to develop new technologies for the aerospace industry.

7.6.1.2 Managing Technology in a Large Company

One of the reasons large companies are able to thwart competition so easily is that they have realized economies of scale. Economies of scale refer to the efficiencies gained because of mass production. As a result, they can enforce barriers to entry for new products and competitors. They have significant staff and resources dedicated to research and development. Furthermore, they can bring ideas to reality by building prototypes in-house. Economies of scale come at a cost, however. There is an immense inertia that needs to be overcome for a large company to adapt to market change. Most of the market changes come with the advent of a new technology or presence of a competitor that changes the market landscape. Larger companies therefore are known to start spin-off companies to develop new technology. This way they can ensure that the research spin-off has some autonomy, and at the same time, the market image of the company is protected in case the new idea fails.

7.7 Design Structure Matrix and Design Churn

Product development, like any other major undertaking, can be subdivided into tasks. It would be great if these tasks were implementable totally sequentially (better managed) or totally in parallel (least time-consuming).

In reality, it is never that way. Product development is inherently interdisciplinary, and different disciplines are rarely synchronized, sequentially or in parallel. The development process is also affected by many controllable and uncontrollable factors. As a result, tasks *interfere*. Interference could imply

1. The tasks have to be done in a particular order.
2. The tasks create rework for each other.
3. At times no clear sequence or parallelism exists.

A design structure matrix (DSM) is a tool to capture such interdependencies in different tasks of a product development process. It was proposed as an engineering tool by Steward (1981), with many later enhancements by Eppinger et al. (1994) and Smith and Eppinger (1998). DSM requires very specific information about individual tasks within a project. Consider four tasks, A, B, C, and D, related to the *design* of two components (A), their *manufacture* in parallel (B and C), and *assembly* (D). Assume now that during assembly design errors are found requiring minor redesign, manufacturing steps incorporating the redesign, and final assembly. These tasks can be represented in a DSM as shown in Figure 7.7.

The entry of an X in a box represents that the task corresponding to the column leads to or causes the task corresponding to the row to start. By definition, a task cannot create rework for itself; therefore, a dot or a dash is entered for the diagonal entries. It is easy to see which tasks are parallel. Notice that B and C do not feed into each other; therefore, they can be done in parallel. Similarly, since A feeds into both B and C, they cannot start until A is done. The full set of four tasks is not sequential because they create rework for each other. For the whole project to converge, it is important that the rework created by D for A diminishes each time.

7.7.1 Partitioning and Clustering a DSM

Partitioning a DSM is a way to make different tasks sequential or as close to sequential as possible. Sequential tasks are easy to manage and track. In the DSM shown in Figure 7.7, to make the tasks sequential, one will have

	A	B	C	D
A (Design)	–	0	0	X
B (Manufacture)	X	–	0	0
C (Manufacture)	X	0	–	0
D (Assembly)	0	X	X	–

FIGURE 7.7
An example of a DSM.

to find an order of the tasks such that they could be completed in a prespecified order, without repetitions and cycles. For the four tasks shown, there is no way to make them entirely sequential. While tasks interfere in any product development process, they do that in a structured way; that is, the interdependencies are and can be usually modeled. For example, design of a large airliner involves many interdisciplinary teams that need to coordinate their efforts. Clustering DSM is the act of organizing the tasks such that related tasks are "clustered" together with minimum interaction with other tasks. As a result, the clusters, while themselves having many interdependent tasks, can work in a sequential fashion with other clusters. The DSM clusters are usually mimicked in actual design teams so that improvements in cycle time are truly achieved in practice. DSM clustering is computationally expensive and requires sophisticated algorithms. Genetic algorithms have been shown to be very effective in clustering DSMs.

7.7.2 Work Transformation Matrix

When looked at in light of results from linear algebra, DSM becomes a powerful tool to understand the state of the product development process. In the DSM shown earlier, the strength of dependencies was not shown. A work transformation matrix (WTM) considers this information (Smith and Eppinger, 1998). Consider three tasks, A, B and C, which all create rework for each other as follows:

1. Every time A is done until completion, 0.6 rework for B is created and 0.3 rework for C is created.
2. Every time B is done until completion, 0.4 rework for A is created and 0.1 rework for C is created.
3. Every time C is done until completion, 0.3 rework for A is created and 0.5 rework for B is created.

We can represent this information in a matrix form, where an entry represents that the task in the column creates rework for the row. Figure 7.8 shows the WTM for the three tasks, A, B, and C.

	A	B	C
A	–	0.4	0.3
B	0.6	–	0.5
C	0.3	0.1	–

FIGURE 7.8
An example of a work transformation matrix.

Following the treatment in Smith and Eppinger (1998), let $\mathbf{w}_0$ be the vector of work that each task initially entails and **R** be the work transformation matrix. Since the entries in the DSM show percent rework, $\mathbf{w}_0$ is simply a vector of 1's. After one cycle the work vector becomes

$$\mathbf{w}_1 = \mathbf{R}\mathbf{w}_0 \tag{7.2}$$

Now the tasks have to perform work corresponding to the entries in $\mathbf{w}_1$. The vector of the total amount of work $\mathbf{w}_t$ performed is given by the infinite sum

$$\mathbf{w}_t = \sum_{i=1}^{\infty} \mathbf{R}^i \mathbf{w}_0 \tag{7.3}$$

If the largest eigenvalue of **R** is less than 1 (the magnitude of real part), then the above sum is guaranteed to be finite. This implies that the process (at least in theory) will converge. Clearly, a lower triangular WTM points to a convergent process in that the total work will converge to a finite number for each task (diagonal entries are zero). This makes intuitive sense because it points to sequential order of tasks, with no cycles. The total work done for each task can be calculated by finding the sum of the geometric progression as follows, multiplied by the task matrix, **D**:

$$\mathbf{w}_t = \mathbf{D}(\mathbf{I} - \mathbf{R})^{-1} \mathbf{w}_0 \tag{7.4}$$

The task matrix is a diagonal matrix, with the diagonal entries representing the work corresponding to each task. Figure 7.9 shows the amount of work done in each cycle corresponding to each task for the above example.

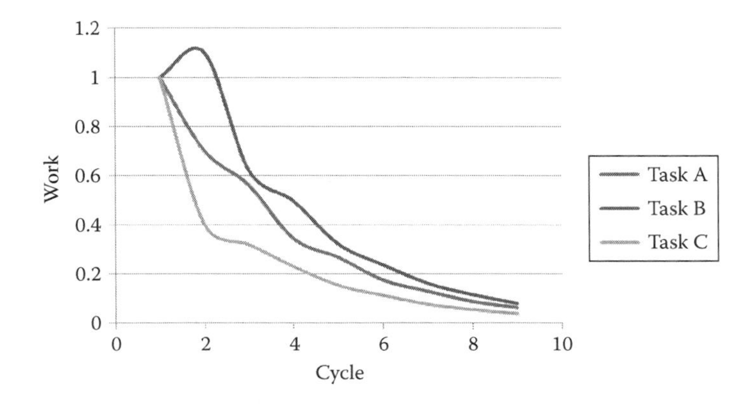

FIGURE 7.9
Amount of work done for each task using the WTM from Figure 7.8.

7.7.3 Design Churn

Yassine et al. (2003) discuss design churn and mitigating strategies for it. Design churn is defined as the oscillatory nature of the product development process in that even a well-managed PD process oscillates between being on time and then running behind schedule. They show that design churn occurs because of inadvertent information hiding every time a "local" change in the system state occurs. This hidden information manifests itself in the oscillations and makes it hard for project managers to understand the progress of the project. Higher-order effects are also present, such as inordinate delays and general lack of team member confidence in the project. Yassine et al. identify three causes of design churn:

1. Different tasks in product design are interdependent.
2. Concurrent tasks sometimes generate more rework than the benefits they provide.
3. There are feedback delays in the state of the system.

Three mitigation strategies are also identified:

1. Identifying tasks that contribute the most to the development time and making them faster
2. Investing more resources to tasks
3. Working on reducing rework

7.8 Product Development and Decision Analysis

In this chapter we covered some of the main topics in product development. The way the two fields stand, we believe that decision based design mirrors product development. The only difference between the two fields is in the perspective chosen. Traditional product development puts the product at the forefront of analysis, while in DBD, a decision maker is the focus. We believe that DBD is a better way to think about product realization because decisions act as essential intervention points in the PD process. Every decision provides an opportunity to make an optimal choice between competing alternatives. This does not imply that traditional product development fundamentals are deficient in any way. Product development has its own formal structure, and because of the advancements in it, we now better understand how to approach the design and manufacture of a product to fulfill customer requirements. Constant evaluation of product design decisions using normative decision analysis can only enhance the quality of the

PD process. It allows for a richer paradigm both in the applied world and in research.

Problems and Exercises

1. What is product development? Why should engineers study product development?
2. How is decision analysis expected to help with product development?
3. What are the challenges to applying decision analytic concepts to product development?
4. Discuss some roadblocks engineers face in thinking about product development.
5. How is creativity from a novice different from creativity from an expert?
6. What is brainstorming? Discuss some pros and cons of brainstorming?
7. How should one stimulate creativity when stuck with a generic design?
8. Estimate demand in Cincinnati, Ohio, for the following products:
   a. A new bagless vacuum cleaner
   b. A 5-inch screen size cell phone
   c. A mechanical pencil
9. What is a functional diagram? Draw the functional diagram for
   a. An MP3 player
   b. An electric guitar
   c. A toaster
   d. A wristwatch
10. Discuss the importance of having a functional diagram during concept generation.
11. What are the two fundamental axioms of design? What is the motivation behind these axioms?
12. How does independence of functional requirements conflict with the least information axiom?
13. Define design for assembly and contrast it with design for modularity.
14. Discuss the technological S-curve.
15. What are all the different design for manufacturing (DFM) tenets?

16. Discuss the role of metrics in a product development process.
17. Discuss mass customization and its impact on economies of scale. Give examples of at least five products that can be mass customized.
18. What are the various ways one can include environmental considerations in design?
19. What is a design structure matrix? What is the eigenvalue condition for knowing that the tasks will require a finite amount of work?
20. What is design churn? What are the causes and potential remedies of design churn?

8

Demand Modeling

8.1 Introduction

The general consensus in the engineering design community is that the main motivation for an engineering firm's activities is to make money. There is nothing necessarily unethical about this notion, though. As Hermann and Schmidt (2006) argue, profit-making enterprises generate employment, manufacture products, and serve the community in the process. Arguably, the contribution to society is greater than the burden, since we do buy their products, which enrich our lives. This decision to purchase also allows the customers to decide the fate of a product, and eventually the company. Of course, the inverse problem of how to make a product that the customers will buy is central to the decision making in a profit-making company. Indeed, there are engineering challenges and regulatory hurdles in making *working* products; the final sales determine profitability and are the driving factors that make a product *successful*. To that effect, demand modeling, the study of the expected number of sales of a product, is undertaken.

Understanding the effect of design decisions on eventual product demand is becoming more and more relevant in today's era of globalization where competitive advantage, whether due to price or technology, does not last long. The focus of this chapter is on understanding how one should think about demand in the context of engineering design. To that end, we will try to understand how to incorporate demand in the initial stages of product design, and what techniques are out there that help predict the expected demand with or without the presence of competition, which will give us an idea of how demand can be integrated into the design decision-making process directly so that the effect of engineering decisions on profit can be readily observed.

8.2 Considering Demand in Engineering Design

We discussed cost-plus and target costing of products in Chapter 7. The cost-plus method evaluates the cost of different design and manufacturing operations, includes logistics costs and a profit margin, and comes up with a price for which the product should be sold. Target costing, on the other hand, requires one to start with a good idea of how much the product should be priced and cascades that down to how much money should be expended on each operation. It is clear that in both cases we should start with some idea of what the price of a product should be; otherwise, we will either end up with a product that is impossible to manufacture or one that no one will buy because it is too expensive. Needless to say, a good model of demand is essential to undertaking engineering design. Furthermore, considering its importance as a major driver of engineering design, demand modeling needs a formal approach.

Assuming that individual customers make rational decisions when purchasing a product, product design and pricing strategies can be devised with a fair degree of confidence, assuming this rationality. It is generally seen that customers lump all the product attributes except price into one,* and then see if the price is justified given the value of other attributes. Product price (particularly in the target costing context) also gives a way to benchmark different designs within a firm and can act as a metric for choosing between different designs. Manufacturers therefore need to know what product attributes are valued by the customers and what cost–attribute trade-off relationship exists in the minds of the customers.

Recall that in Chapter 2, we briefly mentioned Arrow's impossibility theorem, which states that no method of aggregating preferences of multiple individuals can avoid interpersonal comparisons. This applies loosely in a marketplace, as there is no way for customers to rank products available for purchase in the order of desirability with which everybody will agree.† But such ranking is not desired anyway because customers purchase what is most attractive to them *personally*, and the choices they make are independent of each other for the most part. Each customer maximizes his or her own utility over the product choices available. One can also foresee the issue of an optimal landscape of products such that every customer is equally happy with the choices he or she makes. These issues are acknowledged with the convenient assumption that offering enough products with different attributes and functionalities provides an opportunity for each customer to get what he or she wants. The objective of the manufacturer is to get enough

* This is not necessarily in conflict with normative decision analysis.

† If it were true, based on the ranking, there would one unique optimal product that everyone would buy.

customers to buy its product, and also that when it offers multiple products, the distribution of customers over the products is such that maximum profits are realized (different products have different profit margins).

8.2.1 Demand Determines Feasibility

In the initial stages of design, where the manufacturer is evaluating which design to pursue with very limited information, expected demand still plays an important part. In Chapter 2 we presented a decision-based approach to make this meta-decision where the expected utility from the profit (P) is maximized.

$$E[U(P)] = \max_i E[U(R_i - C_i)] \tag{8.1}$$

The revenue term R_i for a design i is the price of the product times the number sold; that is, demand, C_i, is the cost in manufacturing and marketing them, while $U(P)$ is the manufacturer's utility function over money. Large uncertainty at this stage ensures that these values are not known accurately, or even their true distributions. Expert opinion founded on prior experience can be used in making these assessments. For example, a business development manager can provide upper and lower bounds, and most likely a value on how much a product is going to cost and the expected number sold. Demand therefore acts as a feasibility criterion (albeit in conjunction with profit per product) in evaluating new designs. If one design indicates a high likelihood of incurring a loss, or if it is dominated by another design, it can be summarily removed from consideration.

8.2.2 Demand Determines Optimality

Since manufacturers seek only to maximize profit, optimal design should take into account the demand and hence profitability of a product. An explicit mathematical model of demand as a function of attributes, however, is needed. Such a function is embedded into the objective of optimization. More and more new engineering literature incorporates the demand directly into design decision making. Many feasible designs are possible, but few of them are optimal or close to optimal based on the profit criterion. Clearly, one also needs a well-defined mathematical model of the product's attributes given the design decision variables. An optimal product in a profitability sense, therefore, requires the best interplay between its attributes, profit margin, and demand. If there is a way to predict relative demand, optimization can also be used to find what different products to offer in a portfolio versus in a single product. Demand here acts as an optimality criterion, subject to design constraints.

8.2.3 Trade-Off between Accuracy and Ease of Assessment

The reader must be thinking that it should not be hard to fit a mathematical function to expected demand, especially if some data and expert opinion are available. It is an opinion that we subscribe to in this book as well. Indeed, many functional forms and formulations can predict demand to some degree of accuracy; there is, however, generally much more to it. Sophisticated models are generally more accurate but involve parameters that need to be ascertained based on responses from the customers or by observing their previous buying decisions. The ease of assessment of these parameters is what makes one model preferable to another. Simplistic models, on the other hand, are easy to assess, but their accuracy is limited. In general, the preference order over product attributes individually is known. For example, in a car we know that more customers will buy the car if it is cheaper, more fuel efficient, with better acceleration and better safety rating. A simplistic model can be generated that satisfies this preference order. Having a simplistic model like this is very important because involved analysis cannot be performed at each design stage.

The disadvantage of simple methods, as mentioned earlier, is their lack of accuracy. If many product attributes are changed simultaneously, it will be hard to tell whether the demand predicted by such a model can be assumed to be correct. For example, the question "How does the demand change when price is reduced by 10%, fuel efficiency decreased by 5%, and safety rating improved by 3 points simultaneously?" is much harder to answer than when only, let's say, price is changed. Determining when to use an advanced model can be difficult and depends on the time and resources available. In the end, a simplistic method should be given up for a sophisticated model only if the benefit is justified.

8.2.4 Getting the Heuristics Out of the Way

Demand modeling costs time and money. Sometimes it is required to know, with minimal analysis, how the demand of a product is going to be affected, given a slight change in its attributes. For existing products with stable sales this is generally easy to do, particularly if enough prior data are available. As we discussed, luckily the preference order over product attributes is generally known. Here we discuss some simplistic methods that should first be employed before resources are committed to involved methods.

8.2.4.1 Dominance

We studied dominance in Chapter 3. In a deterministic case involving multiple attributes, an option is preferable to another if it is strictly better in at least one attribute, while being at least as good in other attributes. If a dominant product exists, it will be purchased by the customer. This simple concept is sometimes overlooked by many manufacturers when they price their products. An example of dominance in products is seen in the case of consumer electronics. Many small electronic products are sold over the Internet for much cheaper than they are at retail stores since Internet sellers do not have to maintain an actual physical store. As a result, many people buy such products almost exclusively online. Another example is movie rentals, where one has the option to watch unlimited streaming movies online as opposed to going to a video store to rent the movies. There are some straightforward guidelines to follow when it comes to dominance; two basic heuristics are

- Do offer dominating products. Be aware of the competition's offerings and their prices. If it is possible to offer the same commoditized generic products at a lower price, do so.
- Avoid mutual dominance among your own products. Many times car companies have been found guilty of offering two variants where one not only is more expensive, but also has lower fuel economy or suffers on another attribute. Needless to say, the dominant variants will eat up most of the sales.

Dominance does not last for most products because competitors generally start offering similar deals, and in effect, everyone starts offering a similar product at a similar price. Such a scenario has been studied under various market equilibrium models. In the extreme scenario, the manufacturer making inferior products goes out of business. This effect was seen in the solar photovoltaic (PV) cell industry, where research led some companies to make PV cells with higher efficiency at the same cost, leading their competitors to go bankrupt.

8.2.4.2 Simple Reasoning to Deduce Market Size

In the last chapter we briefly discussed using simple reasoning to determine market size of a new product. We revisit it here for continuity in our discussion. Assume that you are the business development manager of a company that manufactures vacuum cleaners. Your design team proposes a bagless vacuum cleaner that is 10 decibels below its competitors in noise level. How will you predict demand for such a vacuum cleaner in a city of 1.2 million people? First, you need to know the number of different households and businesses. If there are 3 people per household, the number of households is 400,000. Now, assuming that each household buys a vacuum cleaner every 4 years, yearly demand for *any* vacuum cleaner is 100,000. Assuming that there are about five competitors, while the lower noise advantage gives you an increase in market share of 25%, the expected demand is 25,000 per year. It is usually seen that market size assessment, by arguing along these lines, provides a pretty accurate estimate. The reason is that the errors in different assessments (competitors, number of households) tend to cancel out. These assessments also allow for a structured way of thinking about demand.

8.2.5 Demand Curve and Price Elasticity

A more formal treatment of the effect of product attributes on the quantities sold is considered under *demand curves*. A demand curve is a plot of demand versus an attribute of concern. For a mass-produced commoditized product this attribute is most commonly price. Figure 8.1 shows a demand curve as a function of price.

Price elasticity is defined as the percentage change in demand, divided by the percentage change in price. Price elasticity is generally negative, since the demand for most products decreases with increasing costs. Exceptions exist, for example, Veblen goods (Veblen, 1899), which see an increase in demand as the price increases because they are considered status symbols.

Price elasticity is mathematically defined as

$$E_P = \frac{\Delta D/D}{\Delta P/P} \tag{8.2}$$

The reason for working with fractional changes is that it makes elasticity independent of the units chosen for price and demand. Note that the current price and demand level should be known for one to be able to define price elasticity. A value of zero for E_p means that the demand is inelastic; that is, demand is independent of the price at the given price–demand level.

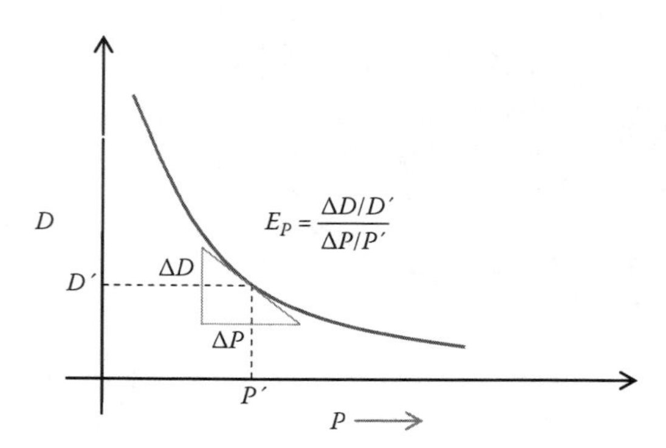

FIGURE 8.1
Demand curve for a product showing price elasticity.

This is rarely true except for very short durations and usually for products that are essential to sustenance. When E_p has a large negative value, even a small increase in price can trigger a precipitous drop in demand, and vice versa. This is true for commoditized products with many competitors. An example is the price of gasoline at nearby gas stations; even the difference of a few cents is enough for most customers to switch to the gas station with cheaper gasoline. Elasticity is greatly affected by demographic factors such as income, presence of competitors, and necessity of the good to sustenance.

Elasticities can be defined for attributes other than price as well. It is a good idea to have a handle on demand elasticities because they give the manufacturer a rough idea of where to expend its efforts. For example, adding more disc capacity to a car's stereo system will not have as much of an effect as the addition of a backup camera, for almost the same cost. The situation might be more subtle in the case of other attributes. For example, when gas prices are high and there is customer sentiment regarding fuel efficiency, buyers might prefer a smaller, fuel-efficient car that may not be as safe as its bigger counterpart. Therefore, expensive additional safety features will show lower elasticity than fuel economy. Elasticities also change with time as customer preferences evolve. Interplay of elasticities over different attributes must also be understood. Consider mobile phones, for example. A larger phone is acceptable only if it comes with a larger screen.

Example 8.1

Acme Industries manufactures semiconductor processing equipment that costs $70,000 to produce. The management considers whether to increase the unit price of the equipment from $100,000 to $110,000. If the expected sales are expected to drop from 10 units per year to 8, should the company go ahead with the price increase? What is the price elasticity?

SOLUTION

Profit before the price change:

$$P_{before} = (100000 - 70000) \times 10 = \$300,000$$

Profit after the price change:

$$P_{before} = (110000 - 70000) \times 8 = \$320,000$$

Therefore, the price increase is justified. Price elasticity is given by

$$E_P = \frac{(8-10)/10}{(110000-100000)/100000} = -2$$

8.3 Demand Models

Demand models are divided into aggregate and relative demand models. Aggregate demand models provide the total demand for a product given the market size and the demographics. Most of the discussion so far in this chapter (albeit simplistic) has been about aggregate demand. In this section we will discuss relative demand models, particularly discrete choice logit models, which are the most common type of relative demand models.

8.3.1 Logit Model

The logit model is a discrete choice model that is helpful in finding the relative demand for competing products. Discrete choice means that instead of directly getting to how many customers will buy a product, we start with a given set of customers and notice which product* they will individually buy *of the given alternatives*, and then extrapolate what we learned to the whole customer base and products with different attributes. The logit model, as a result, is more representative of markets where customers have a choice among different similar but competing products. Even when a manufacturer offers multiple products, the customers choose the one that best satisfies their preferences.

8.3.1.1 Choice Set

Choice set is the collection of alternatives (in our case products) from which a customer is asked to choose. A well-defined choice set is necessary to analyze relative demand of products using the logit model. The model also presupposes that the choices the customer makes are mutually exclusive and exhaustive. Notice that these requirements are not hard to satisfy, as Train (2003) points out. If there are two choices, coffee and tea, one can always define the choice set as having "coffee only," "tea only," and "both tea and coffee" if the customer is allowed to choose more than one alternative. There are also cases where alternatives exist of which we are unaware or do not know the attributes. In such cases, the choice set can be made exhaustive by adding an option "some other product" or "none of the above." It is sometimes referred to as the no-choice option in the demand modeling literature.

The logit method assumes that the value to the customer of a product is given by the sum of two parts: (1) a deterministic or observable part and (2) a stochastic, random, or hidden part. The deterministic part, which we denote as V^o, determines the value derived from the product attributes. This part is assumed to be known or derivable based on survey questions or by

* The choices do not have to be limited to products. The logit model has been successfully used to model relative demands of services, such as transportation choices and even social choices.

observing the customers' choice behavior. The stochastic component, ε, is assumed to follow a distribution, whose statistics may be known but the actual realization is not. Therefore, for the kth customer and for the ith product the total value is given by

$$V_{ki}^{t} = V_{ki}^{o} + \varepsilon_{ki} \tag{8.3}$$

Now, faced with m alternatives, a customer prefers an option i over an option j only if it has higher value, $V_{ki}^{t} > V_{kj}^{t}$ However, the stochasticity in ε's makes the value functions stochastic as well. We can therefore only probabilistically talk about whether the value from one option is greater than that from another; that is, the probability that i is preferred over all other alternatives is given by

$$P_{ki} = P(V_{ki}^{t} > V_{kj}^{t}), j = 1,\dots,m, i \neq j \tag{8.4}$$

or

$$P_{ki} = P\left(V_{ki}^{o} + \varepsilon_{ki} > V_{kj}^{o} + \varepsilon_{kj}\right), j = 1,\dots,m, i \neq j \tag{8.5}$$

We can rewrite the above expression as

$$P_{ki} = P\left(\varepsilon_{kj} - \varepsilon_{ki} < V_{ki}^{o} - V_{kj}^{o}\right), j = 1,\dots,m, i \neq j \tag{8.6}$$

which, in words, implies that option i is better than option j when the realization of the stochastic part is not enough to compensate for the difference in their deterministic values. The above probability can be evaluated if the deterministic parts and the joint distributions of the error terms are known. It turns out that different choices of distributions allow for different methods of calculating the probability. In fact, under restrictive assumptions one can even find a closed-form expression, as we will see. Following the treatment from Train (2003) and McFadden (1974), we rewrite the above equation so that the joint distribution of the error terms comes to the forefront.

$$P_{ki} = \int I\left(\varepsilon_{kj} - \varepsilon_{ki} < V_{ki}^{o} - V_{kj}^{o}, j = 1,\dots,m, i \neq j\right) f(\varepsilon_{k}) d\varepsilon \tag{8.7}$$

where I is the indicator function that is equal to 1 when the logical statement inside the parentheses is true, and 0 otherwise. The joint distribution function $f(\varepsilon_{k})$ provides the likelihood that the expression inside the parentheses is true. If the error terms follow a Gumbel (or extreme value type I) distribution, their pdf and CDF are given by

$$f(\varepsilon_{ki}) = e^{-\varepsilon_{ki}} \exp(-e^{-\varepsilon_{ki}}) \tag{8.8}$$

and

$$F(\varepsilon_{ki}) = \exp(-e^{-\varepsilon_{ki}}) \tag{8.9}$$

We can therefore write the choice probability as, assuming independence between the ε_{ki}'s,

$$P_{ki} = P\left(\varepsilon_{kj} - \varepsilon_{ki} < V_{ki}^o - V_{kj}^o,\ j = 1, \ldots, m, i \neq j\right) \tag{8.10}$$

or

$$P_{ki} = \int \left(\prod_{i \neq j} \exp\left(-e^{-(\varepsilon_{ki} + V_{ki}^o - V_{kj}^o)} \right) \right) e^{-\varepsilon_{ki}} \exp(-e^{-\varepsilon_{ki}}) d\varepsilon_{ki} \tag{8.11}$$

Notice that the $\exp(-e^{-\varepsilon_{ki}})$ term in the integral can be incorporated into the product if we allow i to equal j, $(e^{-e^{-\varepsilon_{ki}}} = e^{-e^{-(\varepsilon_{ki} + V_{ki}^o - V_{ki}^o)}})$. We have

$$P_{ki} = \int \left(\prod_{j} \exp\left(-e^{-\left(\varepsilon_{ki} + V_{ki}^o - V_{kj}^o\right)} \right) \right) e^{-\varepsilon_{ki}} d\varepsilon_{ki} \tag{8.12}$$

or

$$P_{ki} = \int \exp\left(-e^{-\varepsilon_{ki}} \sum_j e^{-\left(V_{ki}^o - V_{kj}^o\right)} \right) e^{-\varepsilon_{ki}} d\varepsilon_{ki} \tag{8.13}$$

The above integral can be easily evaluated using the substitution, $z = e^{-\varepsilon_{ki}}$; applying the limits of integration on the transformed variable, we get

$$P_{ki} = \left. \frac{\exp\left(-z \sum_j e^{-\left(V_{ki}^o - V_{kj}^o\right)} \right)}{-\sum_j e^{-\left(V_{ki}^o - V_{kj}^o\right)}} \right|_0^\infty \tag{8.14}$$

which gives us

$$P_{ki} = \frac{e^{V_{ki}^o}}{\sum_j e^{V_{kj}^o}} \tag{8.15}$$

which is a relatively simple expression for the probability that the kth customer will prefer option i over the other alternatives. Notice that the sum $\sum_j P_{ki}$ equals 1. This is to be expected because all customers are required to choose one (exhaustive) and only one (mutually exclusive) option. Before we solve an example problem to show how logit is used in practice, we discuss the results obtained so far.

8.3.1.2 Properties of the Logit Model

There are some properties of the logit model directly evident from the choice probability expression. The choice probabilities lie between 0 and 1, as one would expect for a probability value. Also notice that the choice probability cannot be zero; in fact, we have to decrease the value from an option substantially to get a value close to zero. If we think that the probability of choosing an alternative is small, we should simply remove it from the choice set. The choice probability follows an S-curve as a function of value from an alternative. This means that the maximum change in choice probability is obtained when the alternative has about 50% chance of being chosen. This is something we observe in real-life purchasing behavior as well. Demand for products that are purchased by very few customers (because of cost, brand loyalty, etc.) or by almost all customers (monopoly) is relatively unaffected when they undergo change in the value of attributes. It is when products take up a substantial portion of the market share (but not all of it) that they are most susceptible to losing out to competing products, or to being well placed for gaining bigger market share, when the attributes are changed.

Another major property of the logit model is called the independence from irrelevant alternatives (IIA). Notice that the ratio of choice probabilities of two alternatives is independent of the value of other alternatives, as shown by:

$$\frac{P_{k1}}{P_{k2}} = \frac{e^{V_{k1}^o}}{\sum_j e^{V_{kj}^o}} \Bigg/ \frac{e^{V_{k2}^o}}{\sum_j e^{V_{kj}^o}} = e^{V_{k1}^o - V_{k2}^o} \tag{8.16}$$

This implies that if we know the values from two alternatives, we can determine their relative market share. Furthermore, this ratio remains constant even if other alternatives are removed from the choice set. While there exist actual products and markets where this substitution pattern is indeed observed, there are many situations where this pattern does not represent reality. Consider a city where commuters choose between a bus, the subway, and personal transportation to get to work. If for some reason the bus choice is not available, we would expect the commuters who usually take the bus to prefer subway over personal transportation. Similarly, introduction of tablet computers and smart phones reduced the dependence on laptops, but it did not really affect the (although shrinking for other reasons) desktop computer markets. In such cases a more flexible substitution pattern is needed. In the next section we will discuss a nested logit model that does away with this restriction. One can always test whether IIA is truly satisfied. As McFadden, Train, and Tye (1978) show, if the parameters of the value functions do not change when assessing them with and without the presence of some alternatives, then the IIA condition is indeed satisfied. Another way to test for IIA would be to include terms from one alternative in the value function of another (McFadden, 1987; Train, Ben-Akiva, and atherton 1989). If these terms have large coefficients attached to them, the IIA does not hold and more flexible methods are needed.

8.3.1.3 Assessing Parameters of the Value Functions in the Logit Model

In the preceding sections we did not really consider the functional form of the observed portion of an alternative's value, V^o. It is generally seen that the logit model is not very restrictive in the kind of functional form that can be used for the value function. Obviously, V^o is going to be a function of product attributes. The way the attributes are combined, though, is given by the functional form and the parameters given by, let us say, the vector $\boldsymbol{\beta}$. Therefore, we can write the probability that the kth customer selects option i as

$$P_{ki}(\beta) = \frac{e^{V^o_{ki}(\beta)}}{\sum_j e^{V^o_{kj}(\beta)}} \tag{8.17}$$

We can use the maximum likelihood estimation method to find $\boldsymbol{\beta}$. A survey is conducted and a set of n potential customers is presented with the choice set. Each customer selects one and only one alternative from the choice set (of m alternatives). The probability that customer k chooses an alternative that he or she was actually seen to choose is given by

$$\prod_{i=1}^{m} (P_{ki}(\beta))^{I(ki)} \tag{8.18}$$

where $I(.)$ is an indicator function equal to 1 when the customer chooses the ith alternative and 0 when he or she does not. In other words, the above expression is simply the probability of choosing the alternative that the customer did end up choosing. Extending this idea to n customers, the likelihood function of observing the choice behavior that we actually observed is given by

$$L(\beta) = \prod_{k=1}^{n} \prod_{i=1}^{m} (P_{ki}(\beta))^{I(ki)} \tag{8.19}$$

The log-likelihood function, which has the same maximum as the above function, is given by

$$LL(\beta) = \sum_{k=1}^{n} \sum_{i=1}^{m} I(ki) \ln P_{ki}(\beta) \tag{8.20}$$

The above log-likelihood function can be shown to be a globally concave function (McFadden, 1974) if the value function V is linear. Recall from Chapter 5 that maximizing a concave function is the same as minimizing a convex function. Therefore, solving for $\boldsymbol{\beta}$, which maximizes the log-likelihood, is a convex optimization problem with a unique global optimum. Its value is given by equating the gradient to 0. We mentioned earlier that the logit model provides reasonable results irrespective of the functional form chosen for v_{ki}. A linear form is therefore routinely chosen because of its simplicity, as well as its ability to find a globally optimum value of the parameters $\boldsymbol{\beta}$.

Example 8.2

In a survey of 50 customers interested in buying a cell phone, the following responses were collected. Only three attributes were considered: screen size (s), battery life (b), and cost (c). Fit a linear functional form for the observed part of the value function, and then find the relative demand for the two choices given by A = {3.3 inches, 5 hours, $240} and B = {3.8 inches, 4.6 hours, $300}.

Choice Number	Screen Size (inches)	Battery Life (hours)	Cost ($)	Number of Customers Willing to Buy
1	3	6	2	20
2	3.5	6	3	10
3	3	4.5	3	5
4	3.5	4.5	2	15

SOLUTION

We use the following functional form for the value from an alternative:

$$V = \beta_1 s + \beta_2 b - \beta_3 c + \varepsilon \tag{8.21}$$

Notice that we use a negative sign in front of the cost term, to ensure that the value goes down as cost increases. This is, however, unnecessary because the optimizer should be able to find this pattern from the choice behavior of the customers. After solving the problem we find the optimum values to be $\beta_1 = 0.4054$, $\beta_2 = 0.3269$, and $\beta_3 = 0.008959$. We will let the reader verify that by using this value function in the logit formula, we get the same choice pattern as the input data set (last column of the table). To find the relative sales of the two given products, we find

$$V_A = 0.822 \quad \text{and} \quad V_B = 0.357$$

And their relative expected demands are given by

$$D_A = \frac{e^{0.822}}{e^{0.822} + e^{0.357}} = 61.4\%$$

$$D_B = \frac{e^{0.357}}{e^{0.822} + e^{0.357}} = 38.6\%$$

8.3.2 Nested Logit Models

Nested logit relaxes the limitation that the substitution pattern applies to all alternatives considered, equally. In nested logit, disjoint subsets of alternatives called nests are defined, and an option shows the IIA property only within its nest. In the previous transportation example, bus and subway will belong to the same nest, that is, public transportation. Similarly, laptops, tablet computers, and smart phones will have the same nest when considering computing and web browsing needs. The stochastic terms in the utility are assumed to follow a generalized extreme value distribution. They are correlated within a nest but not when considering alternatives in different nests.

Assume that there are D nests, with the dth nest having N_d alternatives. Parameter λ_d represents the degree of independence of the stochastic terms within nest d. The choice probability of customer k for choosing an alternative i in nest d, in a nested logit model, is given by

$$P_{ki} = \frac{e^{V_{ki}/\lambda_d}\left(\sum_{j \in N_d} e^{V_{kj}/\lambda_d}\right)^{\lambda_d - 1}}{\sum_{l=1}^{D}\left(\sum_{j \in N_l} e^{V_{kj}/\lambda_l}\right)^{\lambda_l}} \tag{8.22}$$

Notice that the denominator still acts as a normalized factor.

8.3.3 Limitations of the Logit Models

The logit models (both simple and nested) provide a good way to model relative demand of competing products. They have also been shown to predict demand with a fair degree of accuracy in many real-life situations. They both, however, depend on the assumption that the error terms follow some type of extreme value distribution, and they are independent (for alternatives in different nests for the nested logit model). Furthermore, logit models cannot account for random taste variation. Only the observed part of the value functions differs among different alternatives; the error terms are independent of the product attributes. The probit model and mixed logit model can do away with all of these difficulties. However, closed-form expressions of choice probabilities do not exist with probit and mixed logit methods, and simulations must be resorted to. We refer the reader to one of the many demand modeling books for a detailed treatment of these alternative models.

8.4 Using Copulas

Our philosophy in this book is that demand modeling can be achieved through any well-structured approach using a defensible functional form. One interpretation of demand can be drawn from the use of copulas, as shown by Pandey and Thurston (2008). Recall from Chapter 6 that copulas provide a way of modeling joint distributions as a function of marginal distributions. Assume that for each suitably scaled attribute y_i, demand is an increasing function of y_i. If we also normalize demand by the total market size, demand can be viewed as a cumulative distribution function, $F_i(y_i)$. The probabilistic interpretation is that each customer has a cutoff level for each attribute at which the product becomes acceptable. $F_i(y_i)$ is the CDF of the y_i's for all the customers. The joint distribution over all the attributes therefore gives the normalized demand when all the attributes are considered together. Since copulas can combine marginal distributions into the joint distribution, they can be used to predict normalized demand as

$$F_{Y_1,\ldots,Y_n}(y_1,\ldots,y_n) = C(F_1(y_1),\ldots,F_n(y_n)) \tag{8.23}$$

$$\text{Expected demand} = C(F_1(y_1),\ldots,F_n(y_n)) \times \text{Market size}$$

where C is the corresponding copula. Notice that fitting the parameters of the copula can be achieved using a generator function, as in Frank's copula (Nelsen, 2006).

8.5 What Role Does Decision Analysis Play in Demand Modeling?

The decision based design formulation of optimal engineering design will internalize the demand model upon which the decision makers in the company have mutually agreed. The top-level decision is to maximize profit under regulatory and engineering constraints. A firm's sales strategy at this level does not always have to consider the details of engineering design. At this stage the utility from expected profit will be maximized by simply choosing target product attributes. In the design decision-making stage, these product attributes will act as targets to be met. Design optimization can then tell whether these attribute targets can be met, and if not, what is the design that comes closest to meeting these targets.

8.6 Alternative Marketing Strategies

In this section, we go over two alternatives to selling products in a marketplace the traditional way.

8.6.1 The Leasing Scenario

Products do not always have to be sold in the marketplace in the traditional way. Sometimes it is beneficial to enter into a leasing agreement with the customer to provide a service instead of a product. For example, a service may include the commitment from the manufacturer to maintain the product in question in running order by performing routine maintenance. The customer in return pays a predetermined amount at regular intervals. Many such models are already in place, as the examples below show. The benefit of the leasing model is that the demand is predictable, creating a continuous stream of income for the manufacturer. Furthermore, the customer is also locked into a product and will incur a changeover cost if he or she decides to go with a different manufacturer, improving predictability of income for the manufacturer.

Some examples of successful leasing models are

1. Wireless cellular phone plans
2. Photocopiers and printers in many offices
3. Leased automobiles
4. Lawn and farm equipment leases

The predictability that comes with leasing also has sustainability implications, as we touched upon in Chapter 7. Mangun and Thurston (2002) and later Zhao et al. (2010) show how a leasing model of selling products can reduce environmental impacts, increase manufacturer profits, and increase customer satisfaction. Decision analytic methods become even more relevant in leasing type marketing models because products and life cycle lengths can be optimized for the particular customer in question.

8.6.2 Razor and Blade Model

Sometimes a manufacturer gives away a product for free or for relatively cheap because a complementary product can then be sold for profit and with a predictable demand. This idea originated in the razor blade business where a razor was sold for very cheap and customers were charged for replacement blades. Since every manufacturer used a proprietary design, the customers had to purchase blades from a particular manufacturer only. Since blades are a commodity with predictable and almost perpetual demand, this marketing model has been very successful. The approach was later adopted by manufacturers of many different products. For example, ink jet printers are sometimes sold for a loss in hopes that the subsequent purchase of ink cartridges will bring in substantial revenue.

8.7 Demand Evolution and Net Present Value

Demand for products does not always remain the same; it evolves with time and changing preferences of the customers. Moreover, material and production costs also change, affecting the profit margins. Many times demand is influenced by advertisements and the general public's becoming aware of the product. Because most engineering products are designed for a life cycle of many years, decisions made at present should account for net present value (NPV) of the earnings in the future. NPV studies are common in many engineering and finance fields. It is based on the premise that future earnings are less valuable than current earnings. Aside from the uncertainty factor, one justification is that a current amount can always be invested at a risk-free interest rate to recover a future higher amount.

Example 8.3

Consider the following expected demand and profit margin profiles for two product designs. Which should be pursued? (Assume a discounting rate of 5%.)

	Design A		Design B	
Years from Now	Profit Margin ($)	Demand (in 1,000 units)	Profit Margin ($)	Demand (in 1,000 units)
0	50	45	40	60
1	55	85	38	60
2	40	160	38	60
3	30	80	35	55
4	20	30	30	50
5	20	0	30	50
6	20	0	30	50
7	20	0	25	50
8	20	0	25	50
9	20	0	20	50
10	20	0	20	50

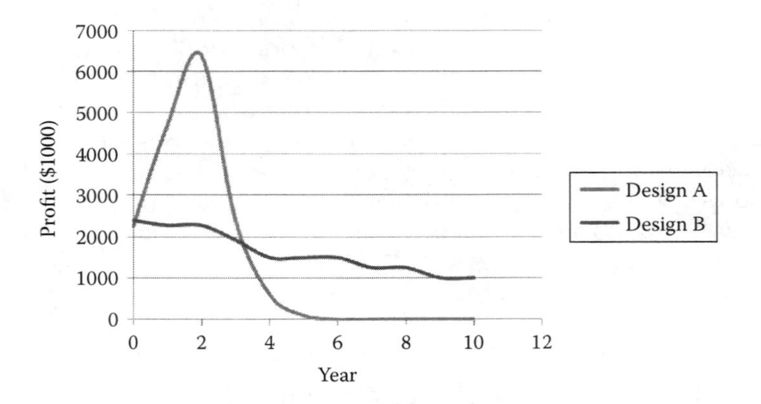

FIGURE 8.2
Profit profiles from the two designs.

SOLUTION

Figure 8.2 shows the profit profiles (profit margin × demand) for 10 years in the future. The profit from design A shows a peak and quickly dies off, while design B provides a more steady profit over a longer period of time. We discount future earnings as

$$\text{Net present value} = \sum_{i=0}^{10} \frac{P_i}{(1+0.05)^i}$$

The NPV value of design A is \$15.07 million, and that from design B is \$14.82 million. Therefore, design A should be chosen. Notice that we did not take into account the uncertainty in the future earnings; if we did, we would have to use a utility function to make this decision. In general, though, earnings this far out in the future (~10 years) are not known with certainty, which makes design B even less lucrative.

8.7.1 Product Diffusion

One of the common ways to understand future sales of products is to study their *diffusion* into the marketplace. The Bass diffusion model is a common technique to achieve that. Frank Bass in 1969 proposed the following differential equation to model product diffusion:

$$\frac{dN}{dt} = \left[p + q\frac{N(t)}{m} \right](m - N(t)) \tag{8.24}$$

where N is the number of customers who own the product, $N(t)$ is the cumulative number at time t, m is the total market size, and p and q are constants to be determined. The term $(m - N(t))$ is the number of customers who do not own the product yet. The Bass diffusion model therefore says that the rate of

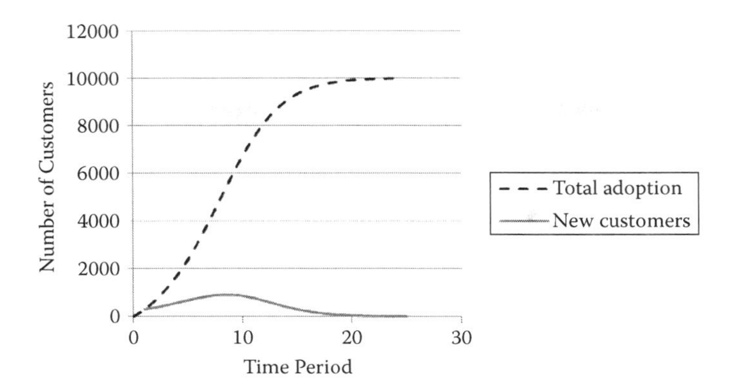

FIGURE 8.3
Diffusion–adoption curve using the Bass diffusion model. The parameters were m = 10,000, p = 0.03, and q = 0.3.

increase in the number of customers purchasing the product at a given time is proportional to the remaining market size, scaled by a factor. The parameter p determines the fraction of the customers influenced by advertisement or external means, while q measures how much the current owners influence the purchasing behavior of the remaining customers. The bass diffusion model and its variants have been successfully used in modeling the sales of many durable products. The values of parameters p and q are therefore known for most common product categories. Figure 8.3 shows the shape of the diffusion curve for a market size m = 10,000, p = 0.03, and q = 0.30. Notice that the total adoption curve resembles an S-curve. This is no coincidence because we have seen in Chapter 7 that technology evolution and adoption cycles follow an S-curve.

Problems and Exercises

1. What is demand modeling? Why should an engineer consider demand when designing products?
2. Discuss and contrast simplistic and sophisticated demand models.
3. Using simple arguments, find the expected demand for the following products in your city:
   a. A generic disposable razor
   b. A generic disposable razor at half the price of the competitors'
   c. An area rug
   d. A lawn mower
   e. A tablet PC

4. What is demand elasticity? How is the price elasticity of demand defined?
5. An increase in $50 in the price of a tablet PC decreases its sale by 20%. If the original price was $400, what is the price elasticity?
6. A lawn mower's price is $800. If price elasticity is –0.4 and the lawn mower costs $500 to manufacture, should the manufacturer increase the price to $900, if the current demand is 3,000?
7. What is the logit model of predicting demand? Discuss some of its properties.
8. Discuss the positives and negatives of the independence from irrelevant alternatives property of the logit model.
9. A survey was conducted to find the demand for a new portable media player. The following responses were collected. Fit a logit model.

Choice Number	Storage Capacity	Screen Size (inches)	Cost ($)	Number of Customers Willing to Buy
1	10 GB	3	250	23
2	20 GB	3	300	12
3	10 GB	2.7	300	7
4	20 GB	2.7	250	8

10. The profit from a product decreases as e^{-t} every year. If the current profit level is $1.1 million, what is the NPV of profits for the next 10 years?
11. How can copulas be used to model aggregate demand?
12. Discuss the Bass diffusion model. What do the parameters p and q represent, give their real-life interpretation.

9

Decision Based Design: Closure

9.1 Putting It All Together

We have covered many topics in this book directly and peripherally related to decision making in engineering design. Engineering design has seen a paradigm shift in recent times where the stakeholders—engineers, business managers, customers, and even legislators—are all considered decision makers in their own right. Each of them makes decisions affecting his or her individual subdomain, as well as those of other stakeholders. Making decisions in such a complex environment requires a formal approach, and normative decision analysis therefore naturally becomes relevant. The perspective we chose in this book was that of the manufacturer who intends to manufacture profitable products by making good engineering decisions, while meeting customer and legislative requirements. These decisions determine the overall attributes of the product, including its profitability.

Massive globalization has ensured that market cycles are shorter and competition stiffer; therefore, engineering design decisions need to be made under severe time, legislative, and market constraints. The highly complex nature of today's engineering design itself has not made this job easier. Clearly, the science of engineering design has kept pace in training engineers in technical matters. There is a perceived gap in the resources available to engineers that will enable them to look at the global picture of successful and profitable product development (PD). In this book we studied the integrative field of decision based design (DBD), which we argue mirrors product development. We contend that decision based design is the best way to realize a product and provides the best perspective on product development by facilitating *good decisions.*

9.2 Decision Based Design

Formal decision analysis is the cornerstone of decision based design. If we are going to consider a decision the basic unit of communication between different stakeholders in a design team, as Lewis and Mistree (1997) argue, we must know how to make good decisions. Decision analysis is a formal methodology for making good design decisions under uncertainty. A good decision maximizes the probability of a good outcome, consistent with the decision maker's preferences. Decision analysis therefore does not question what the decision maker wants; it only helps him or her achieve it with the highest probability. Many researchers in mathematics, economics, and engineering fields have contributed to bridging the gap between the two disciplines of decision analysis and engineering design.

9.2.1 Decision Analysis Fundamentals

Decision analysis is based on strong axioms with which almost everybody agrees. Agreeing with these axioms implies that a rational decision maker maximizes the expectation of his or her utility function, even if it is not explicitly known to him or her. In Chapter 3 we went over one of the ways these axioms have been presented (five rules of actional thought) and also discussed different approaches to assessing utility functions. We also talked about how to frame a decision situation, find the value of information regarding resolution of uncertainty, find the certainty equivalent of decisions, and assess multiattribute utility functions (irrespective of whether independence conditions are satisfied). Both young and experienced engineers should develop familiarity with decision analysis, as it is an important resource in engineering design.

Research in decision analysis is very active. In fact, the increasing complexity of design processes has required that decision analytic methods keep pace with them. To this end, newer methods of assessing utility functions (copulas, Bayesian methods) have been investigated. Faster computers and better simulation algorithms have enabled optimal decision making in many complicated design situations. Recently, it has also been realized that optimization perspectives in integrating decision analysis with design have not been fully investigated. A utility function that reflects decision maker preferences correctly while also facilitating optimization is immensely helpful. Most decision makers are risk-averse, and therefore their utility function is concave. This is very important from an optimization perspective since if we are optimizing on a convex set, utility maximization becomes a convex optimization problem.

9.2.2 Product Development

One of the arguments we have made throughout in this book is that decision based design mirrors product development. In fact, we cannot find any major difference between the two fields in what they aim to accomplish. The only difference we believe is in the perspective. Traditional product development puts the product at the forefront of analysis, while decision based design puts the decision maker at the forefront. We believe that DBD is a better way to think about product realization. This is not to say that traditional product development fundamentals are flawed. Product development has developed its own formal structure, and because of it, we now better understand how to approach the design and manufacture of a product in response to customer requirements. Constant evaluation of product design decisions using normative decision analysis can only enhance the quality of the PD process. In particular, constant feasibility studies using decision analytic methods can help catch errors long before they are able to derail product development. Every decision can be looked at as an intervention point in the PD process, providing avenues for taking the best course of action.

9.2.3 Role of Optimization

Mathematical optimization forms a critical part of design these days. If we have a way of knowing how an engineering system responds to inputs, we can optimize it, that is, find the set of inputs that will give the best-desired output (termed optimal design). Decision based design pays close attention to what function is made the objective of optimization. In the normative domain, it is the multiattribute utility function of the decision maker. Classical optimization techniques depend on closed-form relationships between the inputs and outputs. If this criterion is met, then finding of locally optimal solutions can generally be guaranteed. Otherwise, heuristic methods have to be employed. Many heuristic optimization methods (such as genetic algorithms and simulated annealing) have been developed that can find the optimum or a solution that is close to optimum even for large problems that were previously considered intractable. In DBD an optimal design is the one that maximizes the expectation of the multiattribute utility of the decision maker. Newer optimization methods, in both the classical and heuristic domains, that can find solutions to large engineering design problems efficiently are constantly discovered and developed.

9.2.4 Uncertainty Modeling and Simulation

Most decisions we make involve uncertainty, and engineering design is no exception. Traditionally, most engineering design methods ignored the effects of uncertainty and worked instead with the surrogate of *factor of safety*. Such a simplistic approach not only leads to unreliable, even unsafe designs, but also gives the designer a false sense of confidence about the products he or she designs. Many educational institutions are now realizing how important it is to train students in uncertainty modeling methods. In engineering, uncertainty has a specific connotation—it implies existence of mathematically measurable variables whose stochastic behavior can be understood in precise terms. One way to achieve this is to define probability density functions of the uncertain variables. These density functions are rarely known and need to be assessed. Historically, data were the only input to the equation. The emerging consensus, however, is that the best way of assessing probability density functions is to combine expert judgment with data, as in the Bayesian approach. DBD is alive to uncertainty in design situations, and that is why utility functions that accurately model the preferences of the decision makers under uncertainty are so carefully assessed.

9.2.5 Product Attributes

A product is a bundle of attributes. Most of the information about a product can be captured in a relatively small set of numbers, called attributes. For example, in a car, attributes can be size, fuel economy, and acceleration in a cell phone, these can be screen size, speed of the processor, battery life, and so on. One can even assign numerical values to softer attributes, such as the safety rating of a car and the color of a cell phone. Treating products as bundles of attributes helps us to analyze and select from various candidate designs. The attribute values can be used to convey important characteristics of the product to the customers, while determining whether they would purchase it. They also allow us to define value functions (as in conjoint analysis) and even utility functions (as in normative utility analysis) over them. Most mathematical models of products are aimed at finding the value of the attributes (and hence value/utility functions) given the engineering design parameters, such that simulation and optimization can be performed. As a result, it would not be wrong to say that the objective of engineering design is to find the best attribute combinations in a product, given the constraints.

9.2.6 Demand Modeling

Engineering design is usually undertaken to meet a market need. Profits therefore form the driving force behind engineering design. As we discussed many times in this book, there is nothing necessarily wrong with this notion because profitable companies pay taxes, provide employment, and add value to society. To realize profits, however, a company needs to be able to make products that a large number of customers will buy. Demand modeling does the job of predicting the expected number of customers who will buy a product, given its attributes. Demand modeling does so by observing past customer purchasing behavior or by asking carefully chosen respondents relevant questions. The observations are then used to create demand functions such as in the logit model we discussed in Chapter 8. Being able to predict demand helps companies focus their energies on the right products, and, thus, stay profitable. In such cases, a product line optimization model can directly incorporate the impact of a design decision on the demand and hence, the profit and utility of the manufacturer.

9.3 Closure

Decision based design provides a framework for studying today's integrated and interdisciplinary engineering world. Just as lack of technical knowledge puts severe constraints on our ability to improve or even understand the workings of engineering systems, lack of skills in formal decision making leads to decisions that may not be consistent with our own preferences. The author believes that a common-ground approach to engineering design, something that reconciles experienced designers' opinions with mathematical rigor, is rarely pursued. In DBD, the preferences of the designers and their knowledge are critical inputs in the form of utility functions and design concepts, as well as while modeling uncertainty. In DBD, decision makers and decision making are central. Consequently, many ideas developed independently in engineering literature can be viewed from a DBD perspective enriching both fields. It is hoped that this book will help train engineers, both researchers and practitioners, in how to make optimal and defensible decisions in a variety of design situations.

References

Abbas, A. E., 2009, Multiattribute Utility Copulas, *Operations Research*, 57(6), 1367–1383.

Abbas, A. E., and Howard, R. A., 2005, Attribute Dominance Utility, *Decision Analysis*, 2(4), 185–206.

Arrow, K. J., 1950, A Difficulty in the Concept of Social Welfare, *Journal of Political Economy*, 58(4), 328–346.

Bass, F., 1969, A New Product Growth Model for Consumer Durables, *Management Science*, 15(5), 215–227.

Boothroyd, G., and Dewhurst, P., 1990, *Product Design for Assembly*, Boothroyd Dewhurst, Wakefield, RI.

Boyd, S., and Vandenberghe, L., 2004, *Convex Optimization*, Cambridge University Press, Cambridge.

Clemen, R. T., 1997, *Making Hard Decisions*, 2nd ed., Duxbury Press, Boston.

Dantzig, G. B., 1963, *Linear Programming and Extensions,* Princeton University Press, Princeton, NJ.

Deb, K., Pratap, A., Agrawal, S., and Meyarivan, T., 2002, A Fast Elitist Non-Dominated Sorting Genetic Algorithm for Multi-Objective Optimization NSGA-II, *IEEE Transactions on Evolutionary Computation*, 6(2), 182–197.

Dorigo, M., 1992, Ottimizzazione, Apprendimento Automatico, ed Algoritmi Basati su Metafora Naturale (Optimization, Learning and Natural Algorithms) (in Italian), PhD thesis, Politecnico di Milano, Italy.

Eppinger, S., Whitney, D. E., Smith, R., and Gebala, D., 1994, A Model-Based Method for Organizing Tasks in Product Development, *Research in Engineering Design*, 6(1), 1–13.

Goldberg, D. E., 1989, *Genetic Algorithms in Search, Optimization and Machine Learning*, Addison-Wesley Publishing, Reading, MA.

Gurnani, A., and Lewis, K., 2008, Collaborative, Decentralized Engineering Design at the Edge of Rationality, *Journal of Mechanical Design*, 130(12), 121101-1–121101-9.

Hazelrigg, G. A., 1998, A Framework for Decision-Based Engineering Design, *ASME Journal of Mechanical Design*, 120(4), 653–658.

Hermann, J., and Schmidt, L., 2006, Product Development and Decision Production Systems, in *Decision Making in Engineering Design*, ed. K. E. Lewis, W. Chen, and L. C. Schmidt, 227–242. ASME Press, New York.

Holland, J. H., 1975, *Adaptation in Natural and Artificial Systems,* University of Michigan, Ann Arbor.

Howard, R. A., 1966, Decision Analysis: Applied Decision Theory, in *Proceedings of the 4th International Conference on Operational Research (1966)*, 55–77, Wiley-Interscience. Reprinted in *Readings on the Principles and Applications of Decision Analysis,* ed. K.A. Howard and J.E. Matheson, Strategic Decisions Group, Mento Park, California.

Howard, R. A., 1984, Risk Preference, in *The Principles and Applications of Decision Analysis*, ed. R. A. Howard and J. E. Matheson, paper 34, vol. II, Strategic Decisions Group, Menlo Park, CA.

Howard, R. A., 1988, Decision Analysis: Practice and Promise, *Management Science*, 34(6), 679–695.

Howard, R. A., 2007, The Foundations of Decision Analysis Revisited, in *Advances in Decision Analysis: From Foundations to Applications*, ed. W. Edwards, R. F. Miles Jr., and D. von Winterfeldt. Cambridge University Press, Cambridge.

INCOSE, What Is Systems Engineering?, http://www.incose.org/practice/ what is systems eng.aspx (accessed May, 27, 2013).

Jaynes, E. T., 1957, Information Theory and Statistical Mechanics, *Physics Review*, 106(4), 620–630.

Kapur, K. C., and Lamberson, L. R., 1977, *Reliability in Engineering Design*, 1st ed., John Wiley & Sons, New York.

Karmarkar, N., 1984, A New Polynomial Time Algorithm for Linear Programming, *Combinatorica*, 4(4), 373–395.

Keeney, R. L., 1976, A Group Preference Axiomatization with Cardinal Utility, *Management Science*, 23(2).

Keeney, R. L., and Nair, K., 1975, Decision Analysis for the Siting of Nuclear Power Plants—The Relevance of Multiattribute Utility Theory, *Proceedings of the IEEE*, 63(3).

Keeney, R. L., and Raiffa, H., 1994, *Decisions with Multiple Objectives: Preferences and Value Tradeoffs*, Cambridge University Press, Cambridge.

Khajavirad, A., and Michalek, J. J., 2007a, A Single-Stage Gradient-Based Approach for Solving the Joint Product Family Platform Selection and Design Problem Using Decomposition, presented at Proceedings of the ASME International Design Engineering Technical Conferences, Las Vegas, NV.

Khajavirad, A., and Michalek, J. J., 2007b, An Extension of the Commonality Index for Product Family Optimization, presented at Proceedings of the ASME International Design Engineering Technical Conferences, Las Vegas, NV.

Kim, H. M., 2001, Target Cascading in Optimal System Design, PhD thesis, University of Michigan, Ann Arbor.

Kirkpatrick, S., Gelatt, C. D., and Vecchi, M. P., 1983, Optimization by Simulated Annealing, *Science*, 220(4598), 671–680.

Kirkwood, C. W., 1979, Pareto Optimality and Equity in Social Decision Analysis, *IEEE Transactions on Systems, Man, and Cybernetics*, SMC-9(2), 89–91.

Lewis, K., and Mistree, K., 1997, Collaborative, Sequential and Isolated Decisions in Design, presented at Proceedings of the ASME International Design Engineering Technical Conferences, Sacramento, CA.

Magrab, E. B., 1997, *Integrated Product and Process Design and Development: The Product Realization Process*, CRC Press, Boca Raton, FL.

Mangun, D., and Thurston, D. L., 2002, Incorporating Component Reuse, Remanufacture and Recycle into Product Portfolio Design, *IEEE Transactions on Engineering Management*, 49(4), 479–490.

McCord, M., and de Neufville, R., 1986, Lottery Equivalents: Reduction of the Certainty Effect Problem in Utility Assessment, *Management Science*, 32(1), 56–60.

McFadden, D., 1974, Conditional Logit Analysis of Qualitative Choice Behavior, in *Frontiers in Econometrics*, ed. P. Zarembka, 105–142. Academic Press, New York.

McFadden, D., 1987, Regression Based Specification Tests for the Multinomial Logit Model, *Journal of Econometrics*, 34, 63–82.

McFadden, D., Train, K., and Tye, W., 1978, An Application of the Diagnostic Tests for the Independence from Irrelevant Alternatives Property of the Multinomial Logit Model, *Transportation Research Record*, 637, 39–46.

Mistree, F., Smith, W. F., Bras, B. A., Allen, J. K., and Muster, D., 1990, Decision-Based Design: A Contemporary Paradigm for Ship Design, *Transactions of the Society of Naval Architects and Marine Engineers*, 98, 565–597.

Nelsen, R. B., 2006, *An Introduction to Copulas*, 114–132, 2nd ed., Springer-Verlag, Boston.

Nikolaidis, E., Mourelatos, Z., and Pandey, V., 2011, *Design Decisions under Uncertainty with Limited Information*, CRC Press/Balkema, Leiden, The Netherlands.

Pahl, G., and Beitz, W., 1995, *Engineering Design: A Systematic Approach*, 2nd ed., Springer, Boston.

Pandey, V., Mourelatos, Z. P., and Nikolaidis, E., 2011, Addressing Limitations of Pareto Fronts in Design under Uncertainty, presented at Proceedings of the ASME International Design Engineering Technical Conferences, Washington, DC.

Pandey, V., and Thurston, D., 2008, Copulas for Demand Estimation for Portfolio Reuse Design Decisions, presented at Proceedings of the ASME International Design Engineering Technical Conferences, New York.

Pandey, V., and Thurston, D., 2009, Effective Age of Remanufactured Products—An Entropy Approach, *American Journal of Mechanical Design*, 131(9).

Saaty, T. L., 1990, *Multicriteria Decision Making: The Analytic Hierarchy Process: Planning, Priority Setting, Resource Allocation*, 2nd ed., RWS Publications, Pittsburgh, PA.

Savage, L. J., 1954, *The Foundations of Statistics*, John Wiley & Sons, New York.

Scott, M. J., and Antonsson, E. K., 1999, Arrow's Theorem and Engineering Design Decision Making, *Research in Engineering Design*, 11(4), 218–228.

Shannon, C. E., 1948, A Mathematical Theory of Communication, *Bell Systems Technical Journal*, 27, 379–423, 623–656.

Shupe, J. A., Muster, D., Allen, J. K., and Mistree, F., 1988, Decision-Based Design: Some Concepts and Research Issues, in *Expert Systems, Strategies and Solutions in Manufacturing Design and Planning*, ed. A. Kusiak, 3–37, Society of Manufacturing Engineers, Dearborn, MI.

Sklar, A., 1959, Fonctions de répartition à n dimensions et leurs marges, *Publications de l'Institut de Statistique de L'Université de Paris*, 8, 229–231.

Smith, R., and Eppinger, S., 1998, Deciding between Sequential and Parallel Tasks in Engineering Design, *Concurrent Engineering: Research and Applications*, 6, 15–25.

Steward, D. V., 1981, *Systems Analysis and Management: Structure, Strategy and Design*, Petrocelli Books, New York.

Suh, N., 2001, *Axiomatic Design: Advances and Applications*, Oxford University Press, New York.

Thurston, D. L., 1991, A Formal Method for Subjective Design Evaluation with Multiple Attributes, *Research in Engineering Design*, 3, 105–122.

Thurston, D. L., 2001, Real and Misconceived Limitations to Decision Based Design with Utility Analysis, *ASME Journal of Mechanical Design*, 123(2), 176–186.

Train, K., 2003, *Discrete Choice Methods with Simulation*, Cambridge University Press, Cambridge.

Train, K., Ben-Akiva, M., and Atherton, T., 1989, Consumption Patterns and Self-Selecting Tariffs, *Review of Economics and Statistics*, 71(1), 62–73.

Tversky, A., and Kahneman, D., 1974, Judgment under Uncertainty: Heuristics and Biases, *Science*, 185, 1124–1131.

Ulrich, K., and Eppinger, S., 1999, *Product Design and Development*, 2nd ed., McGraw-Hill/Irwin, Boston.

Veblen, T. B., 1899, *The Theory of the Leisure Class: An Economic Study of Institutions*, Macmillan Publishers, London.

von Neumann, J., and Morgenstern, O., 1944, *Theory of Games and Economic Behavior*, Princeton University Press, Princeton, NJ.

Yassine, A., Joglekar, N., Braha, D., Eppinger, S., and Whitney, D., 2003, Information Hiding in Product Development: The Design Churn Effect, *Research in Engineering Design*, 14, 145–161.

Zadeh, L., 1996, *Fuzzy Sets, Fuzzy Logic, Fuzzy Systems*, ed. G. J. Klir and B. Yuan, World Scientific Press, Singapore.

Zhao, Y., Pandey, V., Kim, H., and Thurston, D., 2010, Varying Lifecycle Lengths within a Portfolio for Product Take-Back, *ASME Journal of Mechanical Design*, 132(9), 091012.

Index

Note: Page numbers ending in "e" refer to equations. Page numbers ending in "f" refer to figures. Page numbers ending in "t" refer to tables.